D0071812
JAPANESE
WOMEN
WRITERS

JAPANESE WOMEN WRITERS

Twentieth Century Short Fiction

Translated and Edited by

NORIKO MIZUTA LIPPIT

KYOKO IRIYE SELDEN

An East Gate Book

M.E. Sharpe

Armonk, New York
London, England

An East Gate Book

This is a revised and expanded edition of *Stories by Contemporary Japanese Women Writers,* Translators and Editors: Noriko Mizuta Lippit and Kyoko Iriye Selden (1982)

Library of Congress Cataloging-in-Publication Data

Japanese women writers : twentieth century short fiction /
edited and translated by Noriko Mizuta Lippit and Kyoko Iriye Selden.
p. cm.
ISBN 0-87332-859-0 (cloth) — ISBN 0-87332-860-4 (pbk.)
1. Short stories, Japanese—Women authors—Translations into English
2. Japanese fiction—20th century—Translations into English.
I. Lippit, Noriko Mizuta. II. Selden, Kyoko Iriye, 1936– .
PL782.E8J37 1991
895.6'30108044—dc20
91-2924
CIP

Printed in the United States of America

ED (c) 10 9 8 7 6 5 4 3 2

ED (p) 20 19 18 17 16 15 14 13 12

To our
mothers and daughters

Contents

Introduction

Noriko Mizuta Lippit and Kyoko Iriye Selden

Readers newly introduced to Japanese literature are often surprised by the extent to which the major classical literary works are the products of women writers. *The Tale of Genji* and *The Pillow Book,* two of the most famous prose works of the classical period, were written by women, as were the majority of the poetic diaries, one of the major genres in classical literature.

The prominent position held by women in classical literature is not limited to prose. The writings of more than one hundred thirty women poets, including many who epitomize the creative voice of the era, appear in Japan's earliest poetry anthology, the *Manyōshū,* which was compiled in the eighth century and includes works by people of widely varying education and class background. The second oldest anthology, *Kokinshū,* which appeared a century later, contains in contrast to the *Manyōshū* only carefully selected works by well-trained poets, works able to meet the clearly stated aesthetic and critical criteria of the editors. Yet here too the works of the women poets match those of their male counterparts in every respect. In fact, such women poets as Ono no Komachi and Izumi Shikibu are generally regarded as the most able and popular poets of the era.

Virginia Woolf once said that traditionally poetry was not a form of expression suitable for women since they could not aspire to place themselves at the center of the universe and to reinterpret it accordingly. The poetic expressiveness of Japanese women poets in the classical period, however, belies this perceptive observation in Japan's case.

The striking creativity of Japanese women in the *Manyōshū* era is undoubtedly associated with the freedom and relatively high political and economic position they enjoyed. Yet in the Heian period (A.D. 794–1185), when women were increasingly relegated to domestic roles under the growing influence of Buddhism and Confucianism, which excluded women from the political and economic sphere, they continued to excel in, and play a central role in the development of, classical poetry.

A reason for this was that poetry came to be defined solely as short lyrical poetry (*waka* or *tanka*), and became the accepted social means of expressing love. Perhaps the most significant factor, however, was that poetry was written in *kana,* the Japanese phonetic alphabet, rather than the Chinese language required for official documents. The kana system, which was commonly identified as *onna moji,* "women's letters," was considered too lacking in sophistication for men to use except in waka poetry. Although men continued to write Chinese poetry in Chinese characters, this poetry became increasingly an intellectual pastime or a means for expressing religious thought, and ultimately it gave way to waka poetry as the mainstream of Japanese poetry.

The development of Japanese prose fiction is closely related to poetry; the major sources of fiction are to be found in the lengthy prefaces to poems and poetic diaries together with the storytelling tradition in the folk and oral literature. The private, lyrical nature of classical poetry was in large measure carried over to prose fiction. Poetic fiction (*uta monogatari*) and poetic diaries, both of which excelled in psychological analysis of the inner world and both of which were written by women in kana, became the major fictional genres in classical literature. Thus Japanese women came to take a leading role in fiction as well as in poetry.

Virginia Woolf's observation that the novel is an appropriate form for women, whom she regarded as especially keen observers of life, is certainly true with regard to Japanese women writers, especially as they became more confined to the private spheres of domestic life, closing the channels for participation in social activities and expression. Yet women's role in storytelling goes back to the ancient period when the mediums, usually women in Japan, were the central transmitters of folk legends and oral literature. In preparing *Kojiki* (712), the first book to appear in Japan (or the earliest which is preserved), a medium named Hieda no Are, who is believed to have been a woman, was assigned to narrate the historical stories of the nation from the mythical period on, while Ōno Yasumaro, a distinguished male Chinese scholar, wrote them down.

In the emergence and development of *monogatari,* a unique fictional genre of storytelling akin to the novel, women writers took a decisive role. In the Heian period, education was the major means of social advancement for aristocrats, and low- and middle-ranking aristocrats in particular placed tremendous emphasis on the education of their children. Although women were excluded from the court examination system, and were therefore exempt from the rigorous studies of Chinese language, history, and literature that men undertook, they too benefited from the emphasis on education.

Since the court hired educated women as governesses and companions to empresses and women courtiers, the demand for educated women provided a considerable incentive for middle- and low-ranking aristocratic families to educate their daughters. The salons of ladies in the court centered around outstanding women poets and musicians who competed among themselves and thus perfected their mastery of art and literature. Yet for the development of monogatari, as for that of waka poetry, the use of the native language of daily life, kana, was the single most significant element. Just as Chaucer's writing in English rather than the Latin of contemporary intellectuals established a new tradition in English literature, the works of women writing in kana became the core of the Japanese literary tradition.

Despite the literary prominence of women, however, their status declined steadily throughout the Heian period, reaching finally the state of complete deprivation of political and economic power and rights which characterized the succeeding feudal era. Women became confined to the domestic sphere and suffered particularly from widespread polygamy. Highly educated in literature and art, sophisticated in understanding the intricacies of human emotions and relations, yet deprived of other social channels for expression, upper-class women recorded in seclusion their sufferings and observations of the life around them.

One of the generic origins of the monogatari, which combines the romance's idealistic vision of love and life with the novel's realistic, critical portrayal of human affairs in society, was the poetic diary, which was typically kept by women. While men kept official, public diaries and historical records in Chinese, women wrote their private thoughts, feelings, and observations of the people around them, developing inner worlds of emotion and psychology in the language of their own daily life. Private, autobiographical writing, profound in psychological understanding, was one significant generic basis of the modern novel in general, but was particularly so in the development of Japanese fiction.

The Tale of Genji, which incorporates elements of epic, tragedy, and lyrical poetry, reveals that women writers had attained a high degree of consciousness of their predicament and confidence in their works. Indeed, the first and among the most influential theories of fiction were developed by its author, Murasaki Shikibu, and by the author of the poetic diary *Kagerō Nikki,* who is identified only as the mother of Michitsuna, a high-ranking court official.

Distinguishing the monogatari from history, Murasaki said that the monogatari deals with human experiences of victors and great people, but also those of weak and socially insignificant people. Dismissing the romances written previously as rank fabrications, the author of *Kagerō Nikki* declared that she was attempting to present her life as it was, to show what the life of the well-to-do woman in Heian society was really like. In this way she made psychological realism the basis of the literary diary as a genre.

The dual structure of Japanese literature, divided between *kanbun* (Chinese writings) and kana writings, is clearly based on the division between the literature of men and that of women. Men did write waka poetry and eventually other literary forms in kana, especially when the aristocracy fell from political power and became nostalgically attached to the Heian literary tradition. Ironically, their copying of the classical works, a reflection of this, helped to maintain the distinct female literary tradition of the Heian period. In the Edo period (1603–1868), when the merchant class became a significant patron of literature, fiction was written in kana; but overall, the male-style literature written in Chinese and in Chinese style never ceased to dominate literary and intellectual writings until the modern period, when Japan turned to the West as a source of learning and freed itself from the influence of Chinese literature.

The recognition of female-school literature as a separate category of writing made the participation of women in literary expression more readily acceptable, creating a situation quite different from that prevailing in the West. Throughout the classical period, educated women continued to write poetry, monogatari, and diaries and confessional memoirs; by the medieval period, and even more so in the Edo period, however, the social system, shaped by neo-Confucianism, confined women absolutely to the home and to the roles of wife and mother. When feudalism was established, women ceased to write, and there ensued a long period of silence in which women lost a major voice in literature. As the *bushi* (samurai) class established its political supremacy based on "masculine" principles, classical female-style literature and expression virtually vanished, and it was not revived *until* the national literature movement in the late Edo period, a movement led by scholars who were dissatisfied with and critical of the Confucian paradigm that stifled literary imagination.

Female-school literature revived fully in the Meiji (1868–1912) and Taisho (1912–26) periods as the pursuit of modernization and the human rights movement stirred in women an awareness of themselves and a desire for expression. Although the modern legal system which replaced feudalism in the Meiji period continued to deny women almost all significant human rights, the introduction of Christian education by missionaries, the arrival of Western feminist thought, and the development of the human rights movement gradually gave rise to outstanding women intellectuals deeply concerned with their status and identity in modern society.

The modernization effort in the Meiji period emphasized the utilitarian aspects of life, relegating literature to play a marginal role, with writers and artists considered basically good-for-nothings. Because writers and artists were ignored and isolated from society, they created a small, closed world of their own, a studio, where they could carry out artistic experiments and live according to their ideas, unhampered by the old conventions and utilitarian concerns. Although they suffered from their isolation from society, it enabled them to pursue and adopt the most radical and avant-garde ideas in the world. Because writers and artists were always in the forefront of the "pure" pursuit of new ideas, and because literature and art were outside the main social activities of Meiji society, these were the most appropriate fields—indeed virtually the only fields—in which ambitious and talented women could seek to achieve self-expression and fame. It was only natural that Seitō (The Bluestockings), which was founded in 1911 and became Japan's foremost feminist group, started as a literary group, publishing the first women's literary journal. Seitō attracted many women whose primary interest was self-expression, and its journal soon became a forum for dealing with all issues related to women and feminism.

During the Meiji-Taisho period, women intellectuals and artists were as preoccupied with the question of the modern self as their male counterparts. Under the influence of writers like Ibsen, the "new woman" gained recognition as a legitimate intellectual subject, although in actuality those women who tried to shed the traditional roles assigned to them and to live freely met social ostracism. Especially characteristic of the literary works written by women during this period is the struggle of the female ego in its pursuit of self-fulfillment and self-expression.

In the modern period too, women writers have been placed in their own group isolated from the main activities of the literary world, and their works have been treated as belonging to a separate category not always regarded as an integral part of Japanese literary development.

This unique phenomenon itself helped encourage women to write because it secured readers and a market for their works. Although Japanese women had historically maintained a relatively high literacy rate, the universal education of women was certainly one of the most remarkable achievements of the Meiji reformation. The recognition of the female school of literature as a separate and legitimate school backed by a long, brilliant tradition and the existence of educated women readers contributed to the flourishing of women's literature by exempting women writers from full-scale competition with men writers and from being subjected to the more acute sexist prejudices in literary criticism which such competition would have engendered. In the same way that separate colleges for women helped promote women's education, the existence and recognition of the separate female-school literature worked as decidedly positive factors in encouraging women to express themselves in literary form.

Yet as more women found the means to express themselves and participate fully in modern life, the tradition of autobiographical, psychological writing associated with the female school became a heavy burden. Limited to dealing with a narrow range of themes and subjects in their private lives, women writers found female-school literature increasingly restrictive. Critics continued to expect women writers to write within their female-school traditions and were prepared to accept only those writers who wrote on such typical "female" subjects as psychology, the mysterious female psyche, motherhood, and female eroticism, subjects which men could not claim as areas of their own expertise. The fact that Japanese literary criticism to this day has been largely a male monopoly must be one of the factors responsible for this.

Okamoto Kanoko, for example, whose writing is characterized by a narcissistic preoccupation with her own female sexuality and her own image as the embodiment of motherhood, virginity, and erotic beauty, is the woman writer most enthusiastically hailed by male critics. Interpretations of her life and works are saturated with the mysticism of female principles, obviously reflecting the aspirations of her male admirers. Furthermore, such general prejudices as the belief that women are not equipped to think originally and in abstraction, that they are too subjective and emotional to deal with broad socio-political issues, and that they are too incapable of comprehending theory to be concerned with aesthetic questions tended to discourage women from addressing themselves fully to these issues.

In addition to the sociological factors that affected the literary assessment of women writers' works, fundamental ideas about women, especially the idea that "female principles" determine the nature of the imagination and thinking of women, exercised a decisive influence in shaping their works. In particular, the "maternal principle," according to which women are supposed to be close to nature and life, has had a great impact on women's imagination and concept of themselves. According to such ideas, women are the transmitters of life, the life-nurturing force in society, and in touch with the mysterious natural and supernatural forces which govern human life.

The mysterious quality of "female nature" or "maternity" has long attracted and fascinated both men and women in Japan. An emphasis on motherhood was only natural in view of the fact that Confucian ethics considered motherhood to be the sole legitimate raison d'être for women, but maternity as a female principle had more than a social dimension. In the modern period, too, the more men and women became wounded and battered from the psyche-damaging struggle of the ego, the more they considered maternity as a refuge, celebrating women with maternal feminine principles as mother-goddesses. The concept of the mother-goddess, however, is only the other side of the concept of women as "witches."

In a culture where Buddhist influence was strong, the socially oppressed sought salvation in the belief in reincarnation and the perpetuation of karma. Ghost stories, including the Noh plays, which form the core of the rich Japanese tradition of the grotesque, gave expression and poetic justice to the silenced, mistreated women. Furthermore, the confessional revelation of the secret self, as in the tradition of *Kagerō Nikki* and in the character of Lady Rokujō in *The Tale of Genji*, tended to reveal the hatred and vengeful desire of the mistreated woman, creating a type of femme fatale. In antithesis to the good wife and mother, there has always been an established image of vicious women, of women as devouring, vengeful, manipulating witches.

Like their Western counterparts, modern Japanese women writers have found the subconscious realm of the female psyche to be a vital area to explore in their attempt to redefine "female principles" and female identity. Writers like Enchi Fumiko and Ohba Minako consider it essential to delve into that area to restore their lost expression. The deliberate use of gothic conventions and the linking of the modern female psyche to the psyches of the mistreated women in the classical period is one of the characteristic features of modern works.

It is true that women writers tended to continue to write in the tradition of female-school literature in the prewar period and even in the contemporary period, expertly exploring the psychology of love, female sexuality, and the emotional intricacies of complex human social relations. Women still are controlled somewhat by the ideal image of women celebrated for so long in the East Asian tradition and even now pervading popular novels and television.

Yet as the work of writers represented in this collection illustrates, affected by and concerned about the events of the age, women did live in history and were subject to the influence of the contemporary political, philosophic, and aesthetic ideas sweeping the nation. Women writers presented a strong voice against war, the atomic bomb atrocities, and class and sexual exploitation. They also wrote about the common predicament of poor, working-class men and women in the capitalist-imperialist period, envisioning ways to liberate them despite the fact that they did not, for the most part, belong to that class. Some committed to socialist revolution debated with utmost seriousness the social function of literature and art and contributed to widening the scope of realism. The works of Miyamoto Yuriko, Hirabayashi Taiko, Sata Ineko, Hayashi Fumiko, Ōta Yōko, and Hayashi Kyōko attest clearly to the social consciousness of women writers.

Other writers write about the possibilities for women in the new life-style, seeing it as a means of freeing themselves from the confines of the traditional socio-sexual roles assigned to women. Questioning fundamentally the conventional ideas about maternity and female principles, these writers risk social ostracism by committing themselves to the pursuit of untraditional ways of living. For some, liberation from traditional social roles and shedding the posture of dependent heroine inevitably bring a confrontation, unmediated by society, with the ultimate meaning of life. Characters in the works of Uno Chiyo, Hayashi Fumiko, Tomioka Taeko, and Murata Kiyoko, for example, face existential questions of the meaning of life as a consequence of their pursuit of a new identity and life-style for their modern female selves.

On the other hand, several writers, including Takahashi Takako and Kōno Taeko, venture into the world of abnormal psychology, violence, and evil in their attempt to fathom the loneliness of modern existence, to explore metaphysically the ultimate meaning of human life. Their world sometimes becomes highly abstract, sometimes fantastic and subterranean—akin to the world of surrealism—and many of their works represent interesting attempts to develop the novel of ideas.

Except for the authors of popular literature, in which traditional women are definitely the most celebrated central female characters, most modern women writers pursue the fulfillment of self and the self-expression which they consider essential to attain identity. Their protagonists emerge consequently as modern female heroes rather than as heroines who are appendages to heroes. They are female heroes who are able and prepared to bear the intellectual and moral burden of reinterpreting the modern world.

In this sense, modern Japanese women writers are not the novelists in Virginia Woolf's definition: novelists who, placing themselves a distance away from full participation in life, write down their observations of people and life. Above all, in their indefatigable pursuit of the ego, they are legitimate and committed participants in the world of modern literature; they emerge from their pursuit sometimes as confident new women, sometimes wounded and battered. Some envision a new social system and an ideal future, while others present a fundamental skepticism toward the fulfillment of the ego in their attempts to transcend the self through religion, madness, or suicide. Their works reveal fully the depth and limitations of modern literature, which has centered on the search for the modern self as its major theme. Their place is assured in modernism and modern literary history.

Despite the full participation of women writers in the main currents of modern Japanese literature, literary critics still tend to treat them separately. Perhaps the most serious shortcoming in the critical treatment of women writers has been the failure to explore the social, historical, and philosophic forces shaping their ideas, imagination, and expression. There has been little effort to evaluate individual authors in relation to naturalism, surrealism, or the other literary movements which are so vital in the case of male writers. Women in general are considered to live outside history, outside the strenuous effort for self-fulfillment that characterizes modern men. Women writers are still today treated separately as a group (with their own associations, literary awards, and so forth) and are evaluated according to criteria different from those applied to men writers. They form at best a separate, unintegrated chapter in the literary history of Japan.

In this volume, we have tried to select works that reveal the range and degree of women writers' participation in modern life and the intellectual and aesthetic development of the modern period in Japan. Our objective is to illuminate their place in the historical development of modern literature, and with this critical intention we have selected works which reveal consciousness and views characteristic of women writers.

Since literary journals constitute the major medium for Japanese writers, authors have generally developed their major themes in short fiction, which is much more significant as a form than in the West, where full-length novels typically occupy the central place in a writer's works. With the exceptions of Hayashi Fumiko and Hirabayashi Taiko, most of the writers in this collection have been extremely prolific and are still writing at a furious pace, in part because of the demand from the journals; this has made choosing stories from among their works quite difficult. Although we have selected outstanding works, in some cases we have chosen representative works over masterpieces in order to illustrate important themes and critical intentions. Furthermore, regrettably we have had to omit many good and significant writers owing to the limitations of space.

Uno Chiyo and Hayashi Fumiko were contemporaries: the former is still writing in her mature years, and the latter died in 1951 at the age of forty-eight. They represent good examples of women writers for whom writing and the choice of life-style were inseparable. They loved and wrote, and many of their major works are autobiographical. Hayashi Fumiko was born in a lower-class family with a peddler as a father and a mother who was even more unconventional in her relations with men than Fumiko herself. Her works are often distinguished by their autobiographical portrayal of a woman equipped with a free spirit and great vitality. Her later short stories, however, including "Narcissus" which appears here, are pure fiction finished with artistic mastery. Uno Chiyo's life was also untraditional in her personal relations, which form materials on which she draws heavily in her works. Yet her treatment of them brings her works close to the phenomenological study of state of mind rather than the involved scrutiny of the self.

Uno's study of human psychology is supported by her stylistic consciousness, especially in non-autobiographical pieces. To quote her from 1977, she sometimes used the approach of "chiseling each letter rather than writing." Uno even invented a special language that would best suit the character by mixing her native dialect with Kansai flavor, most effectively in "Ohan" (1947–50) as well as in some more recent stories. As she is a stylist in literature, she has been a stylist in life as a kimono designer and essayist.

Nogami Yaeko, by contrast, lived a quiet, unspectacular life as the wife of a distinguished scholar of Noh plays who belonged to the literary group gathered around the novelist Natsume Sōseki. Despite her uneventful life, she delved into the central intellectual and moral questions of the day, including those raised by activist movements and by *tenko,* the public recantation of radical political views forced on intellectuals usually through jail and torture, and constantly enlarged the areas of her literary exploration throughout the long duration of her creative activity. She continued to write, mostly in essay form, until her death at one hundred in 1985. Her unfinished last novel, *Mori* (The Woods), a warmly nostalgic semi-autobiographical work portraying a girls' school in Meiji Tokyo, is a monument of the author's mental strength and ever-youthful sentiment.

Hirabayashi Taiko, Miyamoto Yuriko, and Sata Ineko were a vital part of Japan's prewar leftist movements, movements which fought against Japan's invasion of Asian countries and domestic class oppression. Miyamoto, one of the most talented proletarian writers, started writing as a humanist under the strong influence of Tolstoy, but her awareness of women's condition and her own experiences of marriage and divorce led her to become a feminist and communist. "The Family of Koiwai," a good example of socialist realism, traces the growth of a rural, uneducated young girl into a mature woman whose awareness of the general historical and social predicament of women and the proletariat enables her to live with confidence.

Unlike Miyamoto, who experienced fully the censorship and political oppression of writers by being arrested many times yet heroically resisted the pressure for conversion, Hirabayashi was more an anarchist in temperament, and her relation to the movement and Communist Party was indirect, coming through her husband. Nonetheless, she too lived fully in an age of turmoil as a woman and independent thinker who would not be silenced or intimidated by the tenor of the age and the authorities who expressed it. "Blind Chinese Soldiers," because of its controlled realism, is a moving short protest against war; the work depicts not only the atrocities committed against the Chinese but also the victimization of common Japanese people who were thrown into a state of confusion and rendered unable to control their lives. Like Miyamoto, whose sequential autobiographical novels secure her position in modern literary history, Hirabayashi addressed herself more fully after the war to the autobiographical portrayal of her own life.

Sata Ineko also participated actively in the proletarian literary movement, yet after her arrest in 1935, which followed by three years her husband's arrest, she submitted to the pressure to give up her political views and activities. Moreover, she joined the group of writers organized by the military authorities to visit the military front to comfort the soldiers, for which activities she became a target of criticism after the war. Although she rejoined the Communist Party in 1946, several years later she was expelled. "Memory of a Night," written after her expulsion from the party, deals candidly with the theme of wartime collaboration and her disillusionment with the party. As with Uno Chiyo and Nogami Yaeko, Sata Ineko's literary career has spanned the Meiji, Taisho, and Showa eras, giving a special weight to her outstanding reminiscences of the last decade.

Although Enchi Fumiko also started her literary career under the influence of the leftist movement before the war, she went through a long period of silence until she reemerged in the mid-fifties as a writer who delved deeply into the realm of the female psyche. Widely learned in classical Japanese literature, with a translation of *The Tale of Genji* into modern Japanese among her accomplishments, she created spectacular dramas of the female psychic world, connecting it to the one developed by the women writers of the classical period.

The unsublimated frustration of love and the grudges and sorrows experienced by women in polygamy create in Enchi Fumiko's writings a gothic world akin to the world of Noh plays, a world developed with penetrating insight into human emotion and psychology. Her interest in human sexuality is explored in "Love in Two Lives" through her excellent use of a classical work as the structural basis of her story. In "Love in Two Lives" the fictional space is expanded beyond present time and space to portray an archetypal human consciousness of sex.

Ōta Yōko belongs to the generation of Uno Chiyo, Sata Ineko, and Enchi Fumiko, but her achievement is that of a postwar author whose major works probed the world of the atomic bombing and its victims. She wrote both about the immediate disaster and enduring psychological scars from the bombing. She is particularly remembered for her battle against occupation censorship and against her mental illness, as detailed in "Half Human" (1954), caused by the threat of radiation disease and fears of impending world war. "Residues of Squalor" in this volume presents a powerful image of slugs swarming in a dilapidated temporary residence. The narrator sprinkles DDT thinking it will keep them away, but it melts them instantly, reminding her of the A-bomb victims and placing her by analogy in the position of an assailant. Although perhaps unintended, this view of the narrator as both the victim and the assailant, or of society as sharing responsibility for the tragedy, is striking, especially since it dates from the days when, less informed about their own war atrocities, many Japanese focused on themselves as victims. Ōta's effort to record the ravages of the bombing ended with her death in 1963.

Kōno Taeko and Takahashi Takako, writing in the postwar and contemporary periods, also delve into the world of the subconscious and the realm of the female psyche, but both of them develop their dramas in a highly abstract and metaphysical fashion. Kōno's writing is obviously based on subjects derived from her experience, particularly with regard to the inability to conceive, yet her treatment of women's complex feeling toward conception and motherhood, as seen in "Ants Swarm," presents a highly symbolic and metaphysical world in which the condition of human existence is portrayed. In her world, disease, sterility, and the shadow of death deprive people of essential life-energy, and her protagonists are drawn to abnormal psychology to attain a sense of life. Developed under the extreme predicament of death and sterility, her world becomes an allegory of the modern person's quest for life.

Kōno Taeko's female protagonists occupy women's traditional space in a closed room or house, a space in which they pursue sado-masochistic illusions of pain-pleasure sex and child abuse. Their behavior can be understood as a desperate attempt to attain the subjectivity of the female self and sexuality by acting out the masochistic role traditionally assigned to women. In Kōno's works, the outside world exists only as scenery viewed from the window of this closed room.

Takahashi Takako also places her works in a general metaphysical atmosphere of sterility and extreme psychic tension. She has consistently explored the world of madness and evil, delving deeply into the realm of the subconscious, particularly of women. Her recent turning to Catholicism supplies the general philosophic framework for her protagonists' loneliness, fear of life, and the pursuit of the meaning of existence through "sinning." Takahashi consistently deals with the female psyche yet, like Kōno, she presents through her treatment the metaphysical condition in which men and women suffer equally from an inability to attain the sense of life. "Congruent Figures," undeniably a masterpiece, dismisses maternal love as an illusion created by men. It presents maternity, the archetypal fate of female sexuality, as narcissistic self-attachment as well as self-hatred. Both Kōno's and Takahashi's works are among the best examples of the novel of ideas in contemporary Japan.

Although both Tomioka Taeko and Ohba Minako have been writing fiction only in the past two decades, they have already developed their worlds of literature on a large scale. Tomioka, a prolific writer with numerous novels, short stories, dramas, screenplays, essays, and an autobiography to her credit, had established herself as one of the central figures in postwar poetry prior to turning to fiction. "Facing the Hills They Stand," Tomioka's first fictional work, recounts the lives of two generations of a family that settled by an Osaka river; it distinguished her instantly as a superb storyteller. Her attempt to search for the core of human existence, narrated crisply and with ingenious use of dialogue, reveals life's incomprehensible situations in which the only sure things are birth, appetite, and death.

Tomioka has consistently searched for the origins of life in the history of anonymous common people who, without intellectual sophistication and means to escape from life, have confronted life by living it. A deliberately anti-intellectual intellectual, Tomioka rejects overtly the egotistical interpretation of culture presented by intellectuals and artists since the Meiji period, an interpretation which attributes tremendous significance to their own minds and the history of intellectual development. She tries instead to give expression to those silent common people who form the sure undercurrent of civilization. The method of expression she adopts in this story, sometimes referred to as Bunraku-style narration, has been hailed as a highly successful method of presenting the drama of common people. Yet underlying her drama is a feeling for the absurdity of life and an understanding that life must be and is lived by people because it is given to them.

In the 1980s, Tomioka wrote a series of long novels in which she dealt with women who search for a new form of "family," a new human bondage; some experiment with building a women's community, and others try to form a family not based on blood ties and sexual relations. All of them have seen or gone through the trauma of the traditional family and/or the modern nuclear family. Their sometimes comical and grotesque search for a new form of human bonding becomes symbolic of the predicament of contemporary women in the post-feminist era, women who have stepped out of the nuclear family in the hope of attaining their identity as free women, only to be disillusioned with the myth of the "liberated woman" and tortured by the fear of loneliness as well as anxiety over illness and death.

Unlike Tomioka, who has always been inside the literary arena as a modern poet and novelist, Ohba Minako had a long period of submergence as a housewife living in the United States, completely out of touch with Japanese literary circles. During those years, however, in the manner of Jane Austen, she stored observations of life as well as her own frustration, creating a rich reservoir of imagination on which her expression, once it was released, could draw heavily. "The Smile of a Mountain Witch" reveals a devastating perception of human psychology and the logic of people's inner world, presenting traditionally expressionless women as the actual centers of consciousness.

In her recent major novels she searches for the archetypal origins of life and of modern culture, exploring the nature of maternity. She has also written a series of autobiographical novels in which maternity and the search for female identity form the central theme, a theme pursued realistically as well as symbolically in the manner characteristic of her works.

Ohba Minako's search for an archetypal form of life takes her back to the Alaskan village where she had lived for twelve years before returning to Japan and emerging as a novelist. In that remote land's end she finds old friends and their offspring living and dying, remaining together or separated, loving and hating—living in ways common to all human beings, men and women, Japanese and Americans, people of the past and present. She finds their life patterns common to those of all living creatures—animals, fishes, and trees as well as human beings. Lighted by her perception, the human journey through life resembles that of a salmon returning to the river where it was born, there to bear offspring that will travel to the wide ocean, while it dies in its original birthplace.

Hayashi Kyōko emerged as a writer in her forties with a short story dealing with the Nagasaki experience, "Ritual of Death." Although works dealing with the atomic bombing had been written previously, the subject had rarely received a full treatment owing partially to the complex social and psychological issues surrounding the victims. Hayashi herself is a victim of the bomb and her "Ritual of Death" dealt a fresh blow to readers through its detailed autobiographical description of the incident and the weight of the thirty years the author lived with it. Hayashi continues to write on the subject of the atomic bomb both in short stories and in novels, yet her own experience with the bomb, although central, constitutes only a part of the war which occupied her entire childhood. Her experience in China before and during the war, which culminated in 1945 in her return to Nagasaki a few months before the atomic attack, also occupies a central position. "Yellow Sand," included here, is one of the most lyrical and moving short stories to appear in recent years; it deals with the impact that being part of the colonial force had in shaping her mind. In recent years Hayashi has also probed the meanings of marriage and family. Beginning with the examination of her own marriage, which ended in divorce, she has moved on to equally painful observations of her parents' relationship with their children and with each other.

Murata Kiyoko, born in the final year of the war, holds a unique position among those who began their careers in the eighties. She rejects what she calls the physiological/psychological approach used by earlier authors of "I" novels. Instead she is interested in individual origins, which has inspired her to write about old women and the symbols she associates with them, water and earth. Images are important in her writing and often take precedence over plots. As she explains in an interview, she finds an image, usually a tangible object such as a pot, a spotlessly white toilet, or a motorcycle, then develops a story. An author of humor and economy, Murata has a certain reluctance about overly humanizing or individualizing her characters. She avoids, for example, giving her characters realistic names, and when she does she tries to choose plain, psychologically unevocative names. By thus reducing her characters almost to the level of types, however, on the one side she has created an eternal image of grandmother as the uncertain root of life, and, on the other, brought forth a world of adolescents who face unanswerable riddles as they grow.

The works of Japanese women writers since the 1970s can probably be best appreciated when they are placed within the framework of the Western feminist attempts to deconstruct male-centered culture and its subconscious attempts to find a way to release female consciousness from the heavy layers of myth concerning male and female sexuality that have crowded women's psyches and subconscious. Although writers included in this volume are not writing consciously as feminists, their honesty in facing the inner world of their female selves forms unmistakably clear female perspectives. In this sense, their works reflect the major trend of cultural imagination and thought of this century—particularly its latter half—in which women's self-representation and the search for the female self form a major core. Their works represent some of the most penetrating feminist perception, imagination, and thought of the age.

The authors included in this collection of the works of Japanese women writers deal with the experience of modern women with penetrating sincerity and honesty, but their philosophic profundity in understanding modern life, their intellectual capacity to view their experiences in a historical and social context, and their mastery of the art of fiction render the traditional category of "female-school literature" totally inadequate to characterize their works. Indeed, they stand at the core of modern Japanese literature as a whole.

JAPANESE
WOMEN
WRITERS

The Family of Koiwai

Miyamoto Yuriko

I

A night in February—there was not a touch of fire in the room.

Otome, wrapped in a dyed *kasuri* nightcover with a soiled shoulder patch, and leaning her face against the table, sat immobile on a folded sewing board placed across a round, brown porcelain hibachi in which the ashes had congealed.

Severe cold, coming down with the night from the black suburban sky where stars were shining, froze the streets and the earth of the farm fields, pierced the tin roof, and penetrated to the roots of her hair. She felt faintly the warmth of the electric bulb hanging low in front of the table. The electric bulb illuminated the lusterless hair of Otome, sitting near, and the many roughly bound books kept in the beer boxes stacked a bit out of the way at the windowside. The burnish of the table glittered smoothly, its shine so cold that one hesitated to touch it.

Shortly, while keeping her hands across her chest and inside her clothes, and raising only her face from the sleeves of the nightcover, Otome asked her husband Tsutomu, slowly and with emphasis on each word,

"The hot water bottle—is it still hot?"

In front of the same table, Tsutomu, also wearing a nightcover over his shirt in place of a house-jacket, sat on a wicker chair—the only chair

---

Koiwai no ikka (1938). Translated by Noriko Mizuta Lippit with the permission of Kenji Miyamoto. The translation first appeared in the *Bulletin of Concerned Asian Scholars,* Vol. 10, No. 2 (1978).

in the house—resting his cheeks in his palms. Moving his large mouth, conspicuous in the fair-skinned face characteristic of people born in the north, Tsutomu spoke, heavy-mouthed,

"Yes. Shall I give it to you?"

"No, that's all right."

The couple, both covering their small bodies with the nightcovers, almost fell into silence again, but this time Otome, licking her chapped lips anxiously and appearing as if her long eyebrows were raised, said,

"Grandpa may send Mitsuko to us in a package."

"Hmm—"

"Grandpa—we can't tell what he will do."

"—"

On the table lay a letter from Teinosuke, scribbled carelessly with a charcoal pencil on tissue-like paper. Omitting the phrases with which people of the older generation invariably begin, he had written directly to the point, asking when they intended to send the money for which he had written many times. Tsutomu may be engaged in an important movement, although he—Teinosuke—does not know how important, but here a family of five is half-starved. What do you, the eldest son, intend to do? If you do not send the money, then I will make a package of Mitsuko, who is in our way, and send her back to you. Expect that! Teinosuke had cursed the unmatched letters which, sometimes broken, sometimes smudged, reminded them of his stubbornly hairy eyebrows. On the envelope just below the name of Tsutomu—Mr. Koiwai Tsutomu—there was a large oil stain that had penetrated to the letter inside.

In their hometown, A city, Teinosuke had been peddling *nattō* on the street every morning for the past few years. In the evening, Mother Maki, taking the initiative herself, let him take a cart of *imagawa-yaki* sweets to a street on the riverbank which, although particularly windy, was filled with passersby. There they worked until about one o'clock in the morning. Tsutomu's younger brother, Isamu, who had finished elementary school, was working as an office boy at a bank. In this way, the family, including the younger sister Aya, lived.

There was a reason why Tsutomu and his wife had left their three-year-old Mitsuko in the hands of a family so poor. In the spring of the previous year, Tsutomu had been picked up by the police for having worked with a proletarian cultural group, and, because of the beating he had received on his cheeks, he had contracted a middle-ear infection. Tsutomu was moved from the police station to a welfare hospital only when the disease had progressed to the point of causing a brain infection. There a military intern operated on his ear, following instructions to cut here, stuff there, and so forth. The care he received after the operation was so rough as to have astonished a specialist, and in the summer, he contracted quite a bad case of papillitis. Through his friend, he was admitted to a different hospital, but he had remained in critical condition for more than a month. The doctor who headed the ear, nose, and throat department carried out a skillful operation, but even he could not tell for sure whether or not Tsutomu, who lay in bed, a blood-soaked gauze bandage wrapped around his head like a Cossack hat, would recover. As a part of her effort to save Tsutomu, Otome borrowed a kimono from her friend and promptly took it to a pawnshop. Then she took Mitsuko on a night train to her grandparents' place, leaving her there almost forcibly.

It was only for the first two or three months that she was able to send them two or three yen. When autumn deepened, Otome sent Mitsuko a sweater-mantle she had knitted. She was unable to keep the promise she had made to her parents-in-law that she would send them money for rearing Mitsuko.

Tsutomu survived. Since the spring, however, the publishing section of the group for which he had worked devotedly had been beset with great difficulties. There were few who could work, and money was lacking. In the morning, raising the collar of his coat to protect his injured ear, Tsutomu would leave by the front entrance, holding the old essays on Marxism bound in a single volume that he had bought with great sacrifice. After locking the front entrance, Otome would exit by the kitchen door. Receiving several ten-sen coins from Tsutomu at the customary secondhand bookstore, she would return home. This happened more than once.

First her parents-in-law demanded money, and then they began complaining that, with Mitsuko keeping Maki tied up, the volume of sales at her sweets-cart had decreased sharply. They had Isamu write such things for them in great detail. Tsutomu and his wife felt saddened by their inability to send money, yet Tsutomu was annoyed by his narrow-minded father who blamed Mitsuko for the decrease in their business. Nor did Tsutomu, remembering his own experiences at home as a young boy, fail to see the feeling of Isamu, the seventeen-year-old second son, who had had to write that sticky, complaining letter, writing down what his father had said without adding a word of his own. A-city was included in the Tohoku [Northeast] famine district. Since the war started, the deprivation of farmers in that district had been extreme. At the end of the previous year, there was even an incident in which the mothers in families whose young men had been drafted got together and demanded that their sons be returned to them. It was natural for there to have been a decrease in the number of young men who, returning from an evening's amusement, would hold hot baked sweets in the bosom of their kimonos, and eat them one by one. If they had money to spend for such things, they would go to a wonton cart instead of an *imagawa-yaki* cart.

Tsutomu had earnestly explained the world situation in these terms, describing the reasons for their poverty in a way Teinosuke could understand, and writing in the margin that Isamu should be shown the letter too. Teinosuke's letter, which arrived shortly afterward, showed that Tsutomu's effort had been completely in vain. With his dull shrewdness, Teinosuke had begun to use granddaughter Mitsuko in an effort to burden Tsutomu with the responsibility for their straitened circumstances. Moving his disproportionately large mouth as if he were saying "puff-puff," and looking at Otome with sharp eyes, Tsutomu said,

"I never went to a barber before I was eighteen and never bought my underwear." In order to earn money to buy the books which he loved reading, he had worked nightly at a rope factory after his regular job at a post office. His mother, Maki, also found a job at the rope factory. Then she bought a pair of scissors to cut his hair and cloth to sew his underwear, and paid for the medicine for her fragile daughter Aya.

Tsutomu burned the letter from his father. He would be angry, he thought, if his house were searched and the letter seized and used to persuade him to quit the movement.

Otome felt in sympathy with Tsutomu's anger, but, opening wide her eyes with the double-folded lids and looking at her husband's hair which had thinned strangely after the ear infection, she said quietly,

"I hope Grandpa is not mistreating Mitsuko." Otome's voice reflected her dual concern. She felt guilty for not being able to bring up Mitsuko herself, as well as for reminding Tsutomu—already plagued by numerous troubles—of the worrisome family matters.

After the lunar new year was over, Aya, in her unskilled but clear handwriting, wrote to them that Grandma kept saying these days that she wanted to die, and that she worried that Grandma might indeed die. Before Tsutomu appeared the image of the gentle, wise face of his half-graying little mother, carrying the heavy granddaughter on her back, mentally tired and sandwiched between Tsutomu and her stubborn husband. Out of consideration for his mother, Tsutomu agonized over ways to make money to send or to use in bringing Mitsuko back. The poetry writing which had led Tsutomu to become involved in the proletarian movement would bring no money.

It was at this time that the letter with the oil-stain arrived. While Tsutomu, after returning home, sat at the table without a word, Otome scurried about in the kitchen that had no electric bulb, her feet wrapped thickly in Tsutomu's old navy blue *tabi.* She prepared a hot water bottle for Tsutomu. Even the money for charcoal was saved to meet his transportation expenses.

Tsutomu remained silent for a long time. Then, tearing his father's letter with a hand whose middle finger had a red ink-stain, he said in a tone of voice not much different from his usual one,

"I will tell them to close up their place and come to Tokyo."

Otome did not know how to take his statement and looked at Tsutomu as if paralyzed. Then her eyes with two-folded lids gradually grew larger under her unconsciously raised eyebrows, and with the tip of her nose reddened by the cold, she assumed the expression of a startled wild hare.

Closing up the household and the five of them coming to live here—how would they eat? Something akin to fear spread through her and weighed on her. Isn't Tsutomu himself like his father to think of such a thing? she thought.

Tsutomu, however, in between his various activities, had been thinking of this all day. He could not think of any ways to make money, either to send there or to bring Mitsuko back. It was evident that Teinosuke would sink deeper into poverty. In Tokyo, if Isamu worked, Teinosuke sold *nattō,* and Grandma did some part-time work at home using her skillful hands, they would at least be able to eat. Better that they should come to Tokyo and see how Tsutomu and Otome lived. Tsutomu felt certain about it. That would correct Teinosuke's narrow-mindedness, his using the fact that they were taking care of Mitsuko to place the responsibility for their livelihood on the "eldest son" and his letting Isamu's second-son-character develop further. Besides, he might come to understand the nature of Tsutomu's work—understand by being together with them and by seeing the way they lived.

When Tsutomu explained this to Otome, she did not consider it unnatural.

"It may be a good idea."

Thus Otome, her eyes still wide open, gave her wholehearted consent, moving her tongue slowly and moistening her upper and lower lips.

"Then I will send them a letter. You go to bed first."

Tsutomu wrote a letter to Teinosuke and then, taking a long time, wrote something else on a thin piece of paper. Putting each in a different envelope, he took one out of the room and hid it somewhere.

Otome lay in bed facing the *shōji* screen but did not sleep. When Tsutomu told her to go to bed first, she was accustomed to doing so without asking any questions or going into the three-mat room beside the kitchen.

Writing in a small, precise script, Tsutomu often took his left hand out of the sleeves of the nightcover and pressed the wound on his ear with his fingers. The area behind the ear was indented because the bone had been scaled off, and there gauze cloth had been stuffed in. Because of his fatigue and the cold, the wound ached and half his head felt heavy. Behind his ear, in addition to the scar left by the operation, there was a severe scar from a burn. It had been made in the winter of 1930, when Tsutomu, having resigned from the group that was publishing *Literary Front* because he was not satisfied with its direction, had joined the activities of *Battle Flag*. Ishifuji Kumoo of *Literary Front*, a man known for his hunting hat, had placed a hot iron there. It was that scar.

II

Carrying *furoshiki*-wrapped packages of various shapes and colors, Grandpa, Grandma, Aya, Isamu, and Mitsuko all moved wordlessly via Ueno Station to the two rooms under the tin roof of Koiwai. They took up their lives there, spreading even onto the *tokonoma* alcove their sooty packages that contained only rag-like things.

The couple's life changed.

At five o'clock in the morning, while it was still dark, Teinosuke sat up on his mattress and turned on the light above the faces of the family members sleeping next to one another in the narrow room. Preparing his pipe with a noise—"pan, pan"—he started to smoke. Grandma had left out an ashtray for him.

Disturbed by the noise, Tsutomu, who had gone to bed around two o'clock in the morning, turned his body uncomfortably and pulled the cover over his face.

Then Mitsuko started fussing. Otome, who had been patting her daughter's back half-asleep, became wide awake and tried whispering to soothe her lest she wake up Tsutomu. But as if throwing off her mother's soothing words, the bull-necked Mitsuko arched her back, calling for Grandma as she had become accustomed to doing in the past half-year.

"No, Grandma, no."

Tying her apron, Grandma got up from her mattress and said,

"All right, Mitsuko, don't cry. I will give you something to eat."

She brought rice to Otome's mattress and gave it to Mitsuko. Then Isamu got up followed by Aya; Tsutomu could no longer sleep and threw off his thin bedding. While Tsutomu washed his face, dry from lack of sleep, Teinosuke swept cursorily with a broom in front of the entrance and came back to sit in the room, which had already been cleaned and tidied. Aya brought out a dining table. Isamu read the colored advertising leaflet which had fallen onto the tatami from Grandpa's newspaper.

Otome, who was preparing breakfast in the kitchen without cooking utensils, asked,

"Grandma, please taste this," and with the expression formed these days by her raised eyebrows, stretched out a little plate to Maki, who was squatting there. Maki tasted the *miso* soup noisily.

"Seems all right."

Otome and Grandma started serving the *miso* soup, which was diluted and had lots of salt added. Sitting around the small dining table, the family ate breakfast very quickly, and even Mitsuko did not say a word. Even though after breakfast he had time before having to leave for work, Tsutomu did not talk to anyone. Lying on his stomach near the open corridor, he read a book. As if he had just remembered, he would sometimes ask his father,

"What did you do with the tools for baking *imagawa-yaki* sweets?"

"I sold them."

Teinosuke said no more. The conversation between them developed no further. Tsutomu left in a tattered, navy blue coat. Even at that time, Teinosuke never stood up to see Tsutomu off. He remained sitting with his raised shoulders wrapped in a handwoven cotton *haori* jacket.

In the evening, Tsutomu returned. Quite often, Grandpa was still sitting in the same place as in the morning, with the newspaper and ashtray before him. The only difference was that the light was on. While Tsutomu was away, Otome had made and served dumplings out of the flour that Grandma had managed to obtain.

After going to bed in the three-mat room, Tsutomu asked Otome in a small voice,

"What does Grandpa do all day?"

"He sits."

Then, lowering her voice still further as if saying something frightful, she said,

"I hope Grandpa has not lost his senses."

Tsutomu did not answer. Stubbornly sitting thus, Grandpa was watching the way Tsutomu lived. Tsutomu felt it. He knew with bitterness that Teinosuke's mute attitude indicated that he was observing whether the situation dictated that he must work or not.

Every day around five o'clock, Isamu used to go down the gradual slope to a radio shop where he listened to the employment news or looked for a long time into the show window, the blue shade of which reflected the light on his boyish red cheeks. After a month, he was hired by a company near Kyōbashi as an office boy. Tsutomu found for him an old bicycle costing five-yen, twenty-sen, that could be paid for over two months. Isamu commuted happily, pedaling the bicycle, and when he returned at night, he said:

"This company is very big; there are five office boys like me."

He said this with his mouth extended as if he had found out how small the bank in A city was.

"But everyone says about me"—speaking unconsciously in his native dialect—"that although I am young, I am a miser."

At that company, it was a fad among the office boys to treat each other. People said that Isamu was a miser because although he ate when others treated him, he could not return the favor. Grandma, mending the belt of Mitsuko's dress clothes with crudely sewn stitches, worried aloud,

"Then, do not be treated." Tsutomu, for a change, had returned early and was working at the table.

"You do not have to be concerned with such a thing." Moving his large mouth, he spoke gently and encouragingly,

"You can tell them proudly, Isamu, that since you are helping the family you cannot afford to spend money."

Isamu, short, stout, and fair-complexioned—like his brother but with a small mouth—neither protested nor agreed, but began turning the pages of a very old issue of *Children's Science.* Otome wished that Grandpa too would say something at such a time. But he just sat silently, smoking.

However, an incident finally occurred which forced even Grandpa to talk with people and to walk around Tokyo, with which he had not yet familiarized himself. Aya, who was delicate and tended to stay in bed, contracted a complicated stomach infection of tuberculosis. She needed to be hospitalized.

Tsutomu could not return home every night, not only because he was busy but also because his situation had become dangerous. Bringing Grandpa's wooden clogs and making them ready for him to go out, Otome sent him to see his uncle in Mikawashima. Although there was not a great age difference between them, Teinosuke's uncle, Kankichi, had been working as a clerk at a borough government office for over ten years, and was the only relative of the Koiwai family in Tokyo. They had finally come upon the idea of borrowing money from him with the promise to return it out of Isamu's seventeen-yen-a-month salary, and through his government connections, of getting the district committee to be responsible for the medical care of Aya.

Kankichi's third wife, Oishi, visited them after a few days, bringing a contract sheet to receive Teinosuke's seal. The minute Oishi entered through the torn *shōji* screen, she remarked, "What can I do with country folks? They don't even know how to tidy their house."

In an exaggerated way, she walked on her toes in her colored *tabi,* gathering the end of her kimono as if it should not touch the dirty floor. She sat down on the only cushion in the house and looked around indiscreetly at Aya lying in her sickbed. Turning to Grandma who, politely taking off the sashes that tied her sleeves for working, bowed her half-gray head, Oishi said, "Because we are relatives, you use us when you need money, but you do not even show your face in ordinary times. Now stamp the seal here."

Otome raised not only her eyebrows but her skinny shoulders as well. Holding the money Oishi gave to her and carrying Mitsuko on her back, she left to buy ten-sen worth of potato wine and five-sen worth of fried things. Tsutomu had lived for years without having anything to do with his uncle—or with the former-bar-girl aunt whose reputation in their neighborhood was quite bad. As Otome was about to step out, supporting Mitsuko on her back and holding in her hand an empty bottle which might hold about a cup, Oishi said,

"Hey, you. Are you going shopping with that? Even though you want to buy only one cup, they'll give you less than one if you don't take a bottle that contains four cups." Oishi's philosophy of life was like this.

Glued to a spot in front of the table with her eyes wide open like saucers, Mitsuko stretched her hand close to the fried food, almost touching it.

"I want to eat that, Mommy—that I want to eat," she pleaded.

Oishi, protruding her lower lip like a little girl making a face, imitated her,

"This you want to eat, heh?" Looking at Mitsuko with an expression of hatred, she drank wine alone and ate the fried food.

Because you do not believe in religion, you become poor, sickness appears in the family, and your son turns into a red. Using such talk as a relish for her drinks, she finished the ten-sen worth of wine. Belching, she took out another ten sen from her wallet and, putting it inside her obi sash, she sent for more wine. This took place two or three more times until the time came for her husband to return from the office.

When Oishi left at last, Teinosuke took out the debt contract sheet from the drawer of Tsutomu's desk. He looked at it, turning it over again and again. Standing up to put it back again, he said,

"Poverty certainly follows us." It was a tone of voice with deep emotion, a tone in which Otome had not heard Teinosuke speak since he came to Tokyo.

"If one has money, one becomes like that."

"You see, isn't it just as big brother used to say?"

Otome, angry from having to tolerate a woman like Oishi and affected by her expectation that Grandpa was changing his attitude, spoke in a voice that sounded as if her mouth had dried up.

"If society changes, Aya will receive medical care without worry." Licking her lips, Otome explained in detail that in Soviet Russia there is a free medical clinic in each borough to take care of sick people. Tsutomu used to send them the photographic journal *Soviets' Friends* even when they lived in A city. Otome had no way to know what Teinosuke thought when he read it then, but today he certainly listened to her intently. At night she also heard him remark to Grandma,

"I should not have sold the tools for making baked sweets."

III

Turning from the silent main street where everyone was asleep, and bearing to the left around the gas station, Otome entered a newly developed area with only a few scattered houses. The moon suddenly appeared, high and cold, casting its shadow on the ground.

Far and near, *keyaki* trees were enveloped in the haze melting in the moonlight, and in the sky, light white clouds were floating. While walking, looking at the clear moon with the halo around it, Otome felt that only the sound of her shoes and the rustling of her skirt were disturbing the subtle sound of the moonlight falling on everything. It was lonely and frightening to walk alone at such an hour. Yet only while she was walking on such a street could Otome, now commuting to Shinjuku as a bar-girl trainee, regain herself.

Through the efforts of the district committee, Aya was hospitalized in a charity hospital, but the family had to supply someone to take care of her, and that created a transportation expense. Besides, they could not give Grandpa their usual soup-with-everything-in-it as lunch to take to the hospital.

Since Oishi started coming to the family, she began to bring sewing jobs to Grandma and Otome so they could earn some money. They sewed one cotton kimono for twenty-five sen, but they had to supply their own thread. Yet this sewing job was somewhat frightening for them who could not argue well. Early on the promised day, Oishi would come without failure with twenty-five sen.

"I am putting the twenty-five sen here. You work where you are and use a messenger. What high-class people you are."

When Grandma cut the last thread with her teeth and went out to the corridor to dust off the kimono which was just finished, Oishi, who had been drinking potato wine while waiting, examined it immediately.

"Let me see." She folded the kimono as if she were a neat and precise person, and pressed it for awhile by sitting on it. Then, standing up to leave and holding the kimono wrapped in a *furoshiki,* she would extend her hand before Grandma or sometimes before Otome's delicate pigeon-chest, and say,

"Give me twenty sen for I am going to buy tonight's dinner." She spoke looking straight into their faces, without moving an eyebrow. Struck by her aggressiveness and unable to answer, they swallowed their saliva as twenty sen were taken away from the small amount left from what had been placed there.

Otome had decided to work as a bar girl in order to get rid of this witch as soon as possible. But there was another reason as well. It had become apparent that in order for Tsutomu to continue his activities safely, it would be necessary for him to rent a room outside of the house.

Recently, when the printed copies of the journal had been sent to a bindery, he had discovered that the police were after them. At the right moment he took them out quickly. It was such a sudden action that he didn't have anyplace to send them. Therefore, randomly riding around in a taxi, he came to a wooded suburb where he thought to hide the package. It was a Saturday afternoon. As Tsutomu went further and further into the woods, staggering under the weight of the package, he came suddenly upon a narrow, open stretch of lawn. Three young students were lying on the lawn, talking. Both the students and Tsutomu were startled. The students stopped talking. One of them stood up and looked at the short man in a hat who, his large mouth tightly closed, was carrying a package.

Unable to turn back, he continued on and entered the woods that stretched beyond the open area. Deciding upon a certain spot, he began to strap the journals with the paper and cord that he had brought with him. Soon the vibrating sound of a high-pitched whistle came from the direction of the lawn where the students were. It was a jazz phrase with which Tsutomu was not familiar, but he sensed immediately that the hurried tone of the whistle was directed to him and that it was meant as a warning. He covered the journals with grass, placed his coat with its torn collar and hem over them, and listening attentively, pretended to be urinating.

He heard the laughter of women coming nearer and the footsteps of two or three people as they stepped on small branches. When they came to the open area, they seemed undecided but finally turned to the left. The footsteps and cheerful laughter no longer reached Tsutomu.

It took two hours for Tsutomu to finish, and during that time, the students once again let Tsutomu know by whistling that someone was coming into the woods. That night Tsutomu told Otome with deep emotion how the students had conveyed their support. He also told her at that time that he needed a room.

Otome, whose appearance had been changed by the waves newly applied to her hair, used to stand by Tsutomu's table when she returned late at night and, licking her lips that were dry from fatigue, tell him in a low voice what had happened that day at "Beauty Club."

"There is one man—a democrat who sings 'Red Flag.' He showed me a scar on his wrist and boasted that it was from having been tortured."

"Hmm. "

"I felt offended, thinking people will think that communists are all like that."

Tsutomu, who had been suffering constantly from lack of sleep since Grandpa and Grandma had come to live with them, just listened to Otome silently, pressing the wound behind his ear and never asking her about the bar. When Otome talked continuously, he would say sullenly,

"That's enough. Go to bed." He could not get used to thinking of Otome as a bar girl.

As for Otome's suitability for a job, she was certainly an unwaitress-like waitress—more so than Tsutomu or even Otome herself could imagine. One of Otome's customers arrived, sat down heavily in a booth, and gave an order:

"Well, I guess I'll have a cocktail."

Otome, who had been standing and waiting for the order beside him—unblinking and with eyebrows raised—repeated the order:

"One cocktail, right?" She went off with her shoulders raised, repeating the order to herself. She returned with the drink. When the customer extended his hand and tried to touch her, Otome was unable to brush him off skillfully or respond cheerfully with words; when he held her hands, she tightened her body and wordlessly raised high her eyebrows. Her face, with some beauty in it, suddenly assumed the desperate expression of a startled hare. With the unexpected change, the customer felt ridiculous or abashed, and, letting her hand go, reassumed his usual demeanor—but clicked his tongue, "Tut."

After working for twenty days, including the training period, Otome was fired from "Beauty Club." The reason was that, even after so many days, she had not learned how to serve properly. Bringing a book even to bed, Tsutomu asked for the first time,

"What must people do to serve properly?"

"I don't know!" Otome looked depressed, shaking her waved hair.

The way she said "I don't know!" with such strength reminded him of the two of them four years earlier. In the outskirts of A city there was a butcher who had pasted on a glass door a notice "We carry pork." Otome was the daughter of the butcher. When Tsutomu's cousin was hospitalized in A city for a serious eye disease, Otome was working there as a helper. She and Tsutomu gradually started to speak to each other, and Tsutomu, who was working for the post office, lent her *War Flat* and other books to read. Otome had finished only elementary school, but she read it carefully and showed great interest. She borrowed many books from Tsutomu and at one time asked to borrow Marx's *Das Kapital,* possibly mistaking the book for something else. About five days later, Otome, who still had her hair braided, came to the patient's room with drops of perspiration on her nose to return the book.

"Did you understand?"

Tsutomu had unintentionally relaxed his mouth as he asked the question. Otome, raising her long eyebrows so high they might have come out of her forehead, had looked up at Tsutomu who was almost twelve centimeters taller than she (although both of them were quite short), and had said: "I don't know." She had said it with all her strength, shaking her head, just as she did now.

Tsutomu had almost forgotten that, when they were about to marry, he had insisted that he should not have to present the family with the traditional preparation money because he was not accepting a horse or a cow. Otome's mother had cried, saying that since she was not born a horse or a cow, she would like his parents to present a formal marriage gift. Otome said that if her family did not approve, she would leave the house. Then she joined Tsutomu who was already living in Tokyo.

The money that Otome had earned with such great pain—disliking the job and knowing Tsutomu's silent dislike of it—was gone after she returned ten yen to Oishi and paid for Tsutomu's room.

After Tsutomu had moved but before Otome had any time to relax, the debt to Oishi had doubled. Aya had died. They did not have money for the funeral. From whom other than Oishi could the family of Koiwai have borrowed?

Grandpa and Grandma (who held Mitsuko on her back with a sash) returned from the crematory with Aya's bones. Grandma moved some of the packages to the back of the closet which left the front part empty. There she spread a blue, Fuji-silk *furoshiki* wrapper and placed the urn containing Aya's bones. Otome was working the early shift at a bar where she had recently begun to work. She returned early and found Grandma sitting before the urn with Mitsuko on her back, her legs dangling down.

"We do not need red cloth any longer," she said quietly without taking her eyes off the urn. She sighed, "Ahh." Isamu also returned and, without sitting down, looked with embarrassment at the urn. Then he bowed abruptly, lowering his head awkwardly.

No one cried. Mitsuko, her hair in a Dutch cut, turned toward Otome and said repeatedly, "O! O!" pointing at the urn as if she were sorry that it was not something to eat. There was no longer a homeland for Grandpa and Grandma—neither a place to bury her nor a temple to which they could take her bones. The stubborn attitude that Grandpa had assumed since coming to Tokyo gradually began to disappear. Otome could sense it even from the way Grandpa sat.

Since their debt had increased, Oishi came to their house every three days. Learning that Oishi was being inquisitive—asking why Tsutomu stayed away from home so frequently and asking for the location of the bar where Otome was working—Otome warned Grandma earnestly with force in her eyes, "Grandma, be careful. You don't know what she would do to us."

Oishi might do anything to obtain money. It would be nothing at all to Otome if she came uninvited to the bar to threaten, but she might do far worse. Otome shuddered with fear. At the time, Grandma just answered vaguely, "Yes, that's true," without showing whether or not she understood. But that night she must have thought over the matter, and in the morning she came to Otome who was doing the wash in a bucket in the kitchen. Holding Mitsuko, who was poking her hands in the bucket and making a mess, she reported to Otome:

"When Aunt came yesterday, she asked whether Otome too was helping the reds."

"You see, I told you. And what did you say?"

"You are working for a bar, so I told her that you are working for a bar."

Worrying about the time when she was not at home, Otome told her, "Please warn Grandpa well too."

These days, Grandpa had begun to use the baby carriage to peddle sweets to children, going out whenever the weather was good. In the evening, Grandpa would return to the area near home at almost the same time as Isamu. Coming in through the back alley without the carriage, he would check to see that Oishi was not there, and then push the carriage to the front entrance. Once he had bumped into her and she had taken ten sen out of his small sales. It was a bitter experience for him.

At the sound of the carriage's front wheels being raised so that it could be taken inside, Mitsuko, hearing it from somewhere, came rolling out without fail.

"O, o, Grandpa."

She raised her strong-willed forehead in an expression that was comical but filled with happiness.

"I want some sweets."

Sitting with her little legs folded kite-like, she placed her dirty hands on a square package wrapped in a *furoshiki.*

"Hey, wait until Grandpa enters the house."

"No, these are mine."

Grandpa sat down silently just inside the entrance and let Mitsuko have a few cookies coated with sugar. Mitsuko, glancing up at Grandpa and Grandma in turn, quickly put all the cookies in her mouth.

It was about five or six days after the conversation in the kitchen. Carrying both the crochet-work she had knitted while sitting at the "Lily of the Valley" bar—on the worst occasions she was called to work only once in three days—and the sixty-odd sen she had earned, Otome was walking leisurely up the slope toward home when she saw a policeman coming toward her. There was only one road, on one side of which was a cedar nursery. As Otome walked slowly up the slope watching him, he moved from one house to the next checking their name plates. He stopped in front of the house that had a paper on which was written "Koiwai." Opening the door, he entered the house and called in a loud voice,

"Hello, is anyone home?"

Otome's breathing became rapid—not only from climbing the slope —and she unconsciously opened her mouth and looked around. Appearing unconcerned, she turned in two houses before her own house and went around to the back. Grandma had taken out the bucket and evidently had been hanging the laundry. Trying not to make any noise, Otome listened to the conversation in front while wringing out the washed clothes and hanging them on the clothesline.

"The family, are there five now?"

"That's right."

"That child—Ah, it is Mitsuko."

The policeman was silent and it seemed that he was checking his notebook. Soon he shifted his weight to his other leg, making a clashing sound with his sword as he did so.

"So the son, Tsutomu, is missing, huh?"

Holding Mitsuko's little pink underwear, Otome felt as if her ears were filled with sound. Grandma answered in her usual slow, low, and polite voice,

"Yes."

"Why did he leave the house—since he had the child?"

"—"

"Dissipation?"

"Well, that is about right."

Unconsciously, Otome almost smiled. Looking down with tightened shoulders, she said to herself, "Good work, Grandma." She really felt so.

Living in poverty for over thirty years, Grandpa had managed to exist until today without even knowing how to cook rice, and he had managed to put Isamu through elementary school. Sometime in the past Tsutomu had said that Grandpa's life depended on Grandma. At a time like this, Otome sensed Grandma's earnest quick wits.

For the next few days, under the dusty red and green lights of the "Lily of the Valley," Otome recalled vividly the two voices in the conversation, "Dissipation?" "That is about right." Yet the more she recalled it, the more complex became the emotions that accompanied it. The fact that Tsutomu was the very opposite of a dissipated person made the conversation with the policeman indescribably humorous, and as his wife, Otome even enjoyed it. However, the more she thought seriously of his strong character, the more deeply she came to feel about their relationship. She had not thought about it previously. Because of the pressing circumstances, Tsutomu had begun to live apart from her without giving her time to think over their separation. But Tsutomu was not a man who would desert her out of dissipation. She had carried the thought only to this point before. If she did not keep up with his activities in the movement, however, he would not have her as his wife. Now she understood it firmly. If indeed that were to happen, Otome knew that she could not cling to Tsutomu and show him her shame. The value of the proletarian movement and the value of Tsutomu had permeated her being. Thinking of these things for the first time since Tsutomu had left the house, Otome could not sleep for a long time; leaning her face on the table, she held Mitsuko.

It became the season for wearing serge. Frequently a bright, fine rain fell. On the rainy days, the fragrance of the resin from cedar saplings floated subtly from the cedar nursery out front into the open corridor of the two-room house in a closet of which was the urn containing Aya's bones.

It was a late-shift day, and Otome was reading at Tsutomu's desk. Grandpa, who could not go peddling because of the weather, had been reading the newspaper for a long time. "Oh," he called to Otome, taking an unlit pipe from his mouth. "They can't be arrested like that."

"What is it?"

Wondering what it was about, Otome went to look at the newspaper. In the corner of the third page, there were a few lines reporting that two "all union" workers had been arrested at an employment agency in the Kōtō area.

Grandpa's way of reading the newspaper had changed. That was clear to Otome. Grandpa was asking questions which even Isamu did not ask. After listening silently to Otome's faltering explanation, he coughed and, as if thrusting a stick, asked,

"Isn't there a union for peddlers of sweets?"

Otome was flurried without knowing why. She raised her eyebrows and answered,

"I don't know."

Silent again for awhile, Grandpa bit his pipe. Abruptly he took the pipe from his mouth and hit the ashtray with force.

"It will be troublesome if the world does not change in the direction which Tsutomu describes."

It sounded as if he meant that it would be troublesome for him personally, but Otome thought that it showed how much progress Grandpa had made.

"Yes, and that's why, Grandpa, you should not say such things as you did the other day."

About a month earlier, when Otome was hanging up Tsutomu's overcoat, Grandpa had turned its tattered insides out and remarked,

"He is no good, a man nearing thirty, living in Tokyo, and walking around in such a thing."

Otome had involuntarily lost her temper and quarreled with him. Now she was referring to that incident.

Grandpa silently shook his knees, slowly puffing the smoke from his pipe toward the out-of-doors where the rain was falling softly.

Otome soon stood up to change, letting Grandma take care of the clinging Mitsuko. While tying her *obi*, she felt as if she could see Tsutomu before her eyes, clad in a suit, walking along steadily with his mouth tightly closed, and holding an umbrella over his small body.

The Full Moon

Nogami Yaeko

Since it was a beautifully clear, warm day, so unexpected in early February, I had the hairdresser Omiyo do my hair on the second-floor porch. To speak plainly, since I no longer care about hiding my age, I was having her pull my white hairs. I had started to have her over once every month or two more than ten years ago. Thanks to this, I have lived without knowing the great trouble that otherwise accompanies a woman's personal care after middle age.

"Don't ever dye your hair," my mother had told me in her dialect, explaining the trouble she had gone through as a result of starting to dye hers early. She added, "Once you dye it, you can't stop. You suffer for the rest of your life."

I was lucky enough to be able to heed her advice, thanks to Omiyo. As the billboard hanging outside promised—it pictured a young bride in the *takashimada** hairdo, a kimono with flower patterns around the bottom—Omiyo was an excellent hairdresser. Yet she kept her business small, with no disciples or equipment for permanents. She lived with her husband and her mother-in-law, who was good at sewing. All three were Catholic. In addition to being a member of such a trustworthy family, she lived just around the corner, so I could run to her place to ask for help in case of emergency, whether it was house-sitting or household help, not just hairdressing. Because of such a relationship between us, plucking white hairs was also very relaxing for her. She started in the

---

Meigetsu (1942). Translated by Kyoko Iriye Selden with the permission of the author.

*A traditional hairstyle with a high bun in back, worn by formally dressed young women.

morning, ate lunch with us, took a little rest with tea at snacktime, and continued till the late afternoon.

"How patient women are when it comes to the matter of personal care," my husband Shun'ichi complained, impressed.

"There's a phrase, 'three thousand feet of white hair.' I'm having three thousand hairs pulled, so it can't be done in a second. But then I don't have to have grisly hair like a mixture of salt and black sesame. Isn't that right, Omiyo?"

I didn't like a hair turning half white because it looked nasty, especially for people with a lot of hair. I always thought that if my hair was to turn white at all, I wanted it to turn pure white all at once, like the overnight snow. During my journeys in foreign lands, if I saw an elderly woman dressed all in black and crowned with shiny silver hair, I thought how elegant and beautiful, struck by her more than by younger people.

Parting my hair with the thin end of the comb, Omiyo pulled the white hairs, each about an inch long, working gradually from the middle to the side of my head. That day, while she was thus occupied, I was chatting again about my impressions of women in Europe. It was necessary to talk about something. For me the right amount of stimulus to the scalp was so comfortable, and for her the job was so completely automatic, that both of us would doze off otherwise. Moreover, I had just finished some writing and was enjoying a rare feeling of relaxation, which had made me think of having my hair cared for for the first time in a long while. Besides, the warm sun shining on us was also dangerous.

"How they caw," I said lazily.

"They are probably swarming in the Iwasaki woods," Omiyo replied vacantly. I was talking about the crows that had begun cawing a while earlier around the glass doors on the other side.

"There are no longer as many crows around here as there used to be."

"Well, of course, be they crows or sparrows, these days they all know that they'll starve unless they go far into the country, don't they."

We half-smiled together, but even then the crows continued to cry so hard that we thought dozens of them must be flocking around. They sounded as though they were pressing toward some spoil, each poking at it for himself. Yet at times they suddenly became quiet, with just one or two crows crying "caw, caw" in a pitiable yellow voice, each cry separated as if it were cut off. This sounded strangely ominous.

"After all, the crow's cry is unpleasant, isn't it?"

"People used to say that one could tell when a patient at the hospital down there died by the cry of the crows in the Iwasaki woods. When this happened, according to what I heard, no matter how the crows cried, the patient's family didn't hear them, and the cry only reached other people's ears."

Wouldn't it be the opposite? I recalled a dated Kabuki drama, though I didn't remember when I had seen it. In it a woman who heard an ominous cry of a crow goes home—along the walkway to the main stage—feeling worried; there she finds her sick husband killed by someone he had sought to murder. However, my answer didn't reflect this:

"Then, it means that everything's fine at your place and mine, right? Since both of us hear them all right."

To be honest, I had to utter such artificial words because the crow's cry gradually made me nervous, and I felt the spring of my heart, which had been comfortably relaxed from the morning, coil up with a screech. My mother at home, past eighty, and my oldest son S who lived far off in a foreign country—my vague anxiety was probably related to the two of them.

That night we took our third son Hikaru out for dinner. This was a rare event. Besides, the food was good, and the evening air was as warm as daytime, so we came home feeling splendid.

"Was there a phone call or anything?"

To the usual question I always asked at the threshold, the maid answered, "There's a telegram."

Slipping past Shun'ichi who had gone in ahead of me, I opened the sliding door to the dining room with a noise. Side by side on the table with the evening newspaper was the telegram:

"MOTHER'S STENOSIS RETURNED."

"I guess I won't be able to get an express ticket for tomorrow."

As I turned around toward my husband, still holding the telegram, my voice neither trembled nor sounded high or low. Since I had heard the crows cry that morning, I thought I had been waiting to hear this news. Mother, I'll come home right away, my entire being shouted. I never tried to think about "if anything happened before then." That would never happen. I would not let it happen. As I changed in the back room, somehow setting my teeth as though in resistance, tears ran down my nose.

The second telegram arrived at midnight, indicating that if I was coming home, the earlier the better, although mother's condition was not yet critical. Opening wide my eyes that had hardly closed in sleep, I thought of the many mountains and rivers over those five hundred miles. Kyushu is indeed far away—given the situation, the thought weighed on me.

I let a day pass, and the following day I left Tokyo Station by the Fuji, seen off by my husband and third son, who promised to follow soon, depending on mother's condition. In view of the recent confusion in transportation, I was lucky to be able to get an express ticket the day before. No matter how impatiently I might have dashed out, a regular train would have arrived even later, or required an overnight stay somewhere on the way. At Shimonoseki I was met by Murao, in effect the manager of my younger brother's branch store, who had come from Kokura.

"I know how worried you must be."

With a polite friendliness typical of one trained since early youth, he greeted me with words of concern the moment he spotted me. I asked him if there had been a call or something about the most recent development.

"No, nothing particular."

If so, mother's condition must have stabilized a bit. In any case, thinking that it was just one more step till home once I crossed the strait, with Murao carrying my suitcase, I hurried toward the ferry bobbing on the aquamarine waves. We were pushed and jostled in the tremendous crowd, among which there was, for example, a Korean woman in white *chokori,* carrying on her back her baby covered with a quilt.

It was customary for Murao to meet me there whenever I went home, and then, for my younger brother to come meet me at Beppu or some other place. As the slow train which stopped at every single station gradually approached the town, carrying passengers who seemed like visitors to a hot spring, I made myself a bet: if I see my brother there, mother was certainly temporarily better; if not, her condition required that he not leave her bedside even for a minute. When the train glided into the platform with the deep purplish blue bay expanding around to the left, and the mountains puffing up white steam here and there on the slopes near the base, my face met his through the glass. He was standing right alongside the window where I sat. By the time the train started to move again, he was seated by my side, telling me of mother's illness.

"Now I feel I rushed too much, but in any case, I thought it would be better to ask you back just once."

"That's all right. For I want my trip to prove a waste."

"And, big sister, here's what we arranged—"

My brother continued, asking me to pretend that I was stopping by with no knowledge of her illness while visiting my second son in Fukuoka, lest she think her condition so serious that I had to come home. He explained that he had told her that a telegram from Fukuoka had announced my sudden visit home, and that she believed it completely. I nodded in my shawl.

Mother was in bed in my younger brother's soy sauce shop, the "opposite shop," across the narrow street of the country castle town from the main house's wine shop. The main house, locally called the "mother house," had been occupied by my big brother and his wife, our older cousin who had come from the opposite shop about the time my younger brother moved there. The couple had long been dead. And the twenty-year-old man who was their heir, though not our cousin's own child, was confined to a villa as a result of a chest disease. He left both the store and the brewery entirely with his employees. Mother was trying to maintain this old empty wine shop despite her age. That was how she lived. No matter how we tried to persuade her, she never consented to live with my younger brother. She went over only for supper and ate merrily with her grandchildren, but as she put down her chopsticks and enjoyed a few puffs through her old long pipe with the silver mouthpiece, she raised herself saying, "Well, I'll be off." How about a bath or something before you leave, my sister-in-law Hatsuko might say, but mother would answer that a bath was hot and ready in the main house, too, and stand up quickly. It would have been different if since youth she had been a woman of ability with a sharp tongue and equally able hands; but since she had left business matters to her husband and to the manager, believing that it was nothing women should interfere in, she was nothing but a well-meaning, relaxed, retired woman. The habits described above fundamentally served no purpose. It was simply a kind of religious persistence toward the family, a mode of life which could not be changed. It pleased her to be alone without interference, but they took care so that nothing untoward might happen at night which they would discover only in the morning. A woman by the name of Otane, also past seventy, lived with her; she had lived in the city when she got married, and then, divorced and without children, served at an aristocrat's mansion. What Otane was to mother was exactly like the shadow to the form. Also living there was our older sister-in-law's half brother who served as the manager. His wife was also childless. These constitued the "people inside the mother house."

In winter, however, mother more often than not sat in the opposite shop. Wine brewing, which started during the three months of extreme cold, was almost a religious event, and mother habitually mentioned its importance. Even so, she crossed the street more and more often because the rooms in the opposite shop were a few degrees warmer than the spacious main house, which resembled a shrine with its shiny black pillars. Since mother's first attack of stenosis, I was told, my brother and his wife kept her in their shop. The recent attack suddenly occurred when she was resting in the sunroom on the second floor, a rustic

Western-style room with large windows on three sides, facing the yard.

"Mother, I hear you're in bed," I said, deliberately sounding surprised as I opened the sliding paper door. Having moved to the ten-mat Japanese-style room filled with the bright afternoon sun, and lying with her back against the raised floor, mother smiled cheerfully with her chin on the velvet collar of the brown and yellow striped quilt.

"Are they both fine in Fukuoka?"

Mother asked right away, not mentioning her own illness. Thinking of her affection for her grandson and granddaughter-in-law whom she never saw, I found it a little painful to deceive her. So after answering her as simply as possible, I reverted to the topic of her illness: I told her she looked no different than usual, seeing her from outside. This was not a lie. Her white face, on which even in my childhood I had never seen a faint ruddiness or ample roundness, gained a natural tinge of ivory as she aged. No decline appeared in her cheeks, and she looked healthier than when she was young. As she lay resting, she didn't look as if she was suffering from a terribly painful disease. As long as there was no attack, she felt the same, she said. Spending the winter in the Western room over there without catching cold even once, she had gotten sunburned in the warm sun shining from morning till evening, she said, patting her forehead somewhat shyly. I smiled in my heart at mother, who no matter how old she was spoke as if she hadn't forgotten her personal care since her days as a young wife. I said,

"I don't see Otane."

"Ah, yes, Otane is away for a while. She'd rush over as soon as you let her know you're here."

"Is she all right?"

"She's not declining or anything, but she's quite deaf. I can't communicate with her unless I shout."

"That hard of hearing?"

She used the word *orabu,* one of the ancient Japanese expressions still remaining in Kyushu. Imagining the two old women's conversation, mother shouting, or *orabu*-ing, and Otane bringing her ear closer, I smiled. At the same time, I was reminded that mother, though the older of the two, had good hearing, didn't have a single false tooth, saw well, and was normally healthy enough to take care of mending all the grandchildren's socks. I could hardly believe that anything final would develop at this point, and felt as though the chatting by her bed were no different from that at a usual relaxed homecoming.

But there was another attack in the late afternoon. We surrounded her, stroking her back, patting her legs and back as mother groaned, clenching her teeth, a cold sweat breaking out in drops on her forehead, her outlandishly high nose swelling. Meanwhile, the nurse had given her a shot in the right forearm. After about a quarter of an hour her pain gradually faded. I was clumsier than anyone else, whether I pressed or stroked. The most skillful, after all, was my sister-in-law Hatsuko, who was used to handling mother. With the attack coming on, she held mother's back as she pushed away the hot-feeling covers, and kept pressing the hump that rose along the back. It seemed that this helped her somewhat, but I could not easily find the hump. Even the nurse could not press enough, and mother scolded her frantically, "Press harder." She didn't normally raise her voice like this; all the more for this I could sadly imagine what great pain she must be in.

Once the attack passed, however, mother returned to her ordinary self. She not only pleasantly chatted with me, asking about my house in Tokyo and my son in a foreign country, but had her usual healthy appetite: she ate as much as three bowls of rice with her favorite mild-tasting side dishes, whether raw fish, broiled fish, or things boiled with soy sauce.

"Seeing her condition now, it looks as if she could easily recover if only the attacks could be prevented, doesn't it?"

"That's right, big sister. So we're hoping we'll be able to help her become healthy once again—"

"You press skillfully, Hatsu, but it's too much if you have to get up in the middle of the night too."

"Another nurse'll be coming this afternoon. We'll have enough hands then."

Two or three days later, by the time I had had this exchange with my sister-in-law, I too was used to mother's condition. While she seemed quite well during the day, at night she suffered wave upon wave of excruciating pain, and finally toward dawn she slept. After such nights, she looked markedly tired and started to use the bedpan which she had avoided until then. But even though she felt like it, often she passed no water. Of the three physicians who visited her, one diagnosed it as stenosis caused by an aneurysm, and another seemed to share this view. But the physician in charge insisted on atrophy of the kidney, ascribing the difficulty in urinating to that cause. Since she was taking sufficient nutrition, there would be no rapid change unless she caught cold or something, but in any case, past eighty, it would be difficult for her to recover. On this point, the three agreed.

Visitors came in turns. Just as mother had predicted, Otane, too, rushed from her house in the country.

"My, Otsū, so you're home. We haven't seen each other for such a long time, Otsū."

She called me by my childhood nickname; then sidling up to mother's bedside, Otane told her she was surprised to learn that she was sick, adding nevertheless that she must be happy to have Otsū home: "She lives so far away, you can't see her except for this kind of occasion."

Close relatives and old acquaintances who were shown to her bedside proved a diversion to mother as long as the attacks didn't occur. Since Otane made everyone raise his or her voice loud enough for her hard of hearing ears, a somewhat cheerful atmosphere was created. On such an occasion, Otane's age served as a humorous topic.

"When I was eighteen, she was twenty-seven—"

It was always her habit to compare her age with mother's in this way, without telling her current age. Now she repeated this again.

"While you were cheating by a year or two, you lost track of your age yourself, isn't that right?"

Someone like Hatsuko would speak without reserve from friendly intimacy, but Otane could not catch it on the spot. "What did you say? What?" She brought her ear closer then grinned around her mouth where small wrinkles were neatly folded. She was beautiful from youth; besides, she had no children. She seemed as if she could easily cheat by not just a year or two but even six or seven years, passing for whatever age she professed to be; but thanks to that, even mother's age became obscure.

"I think mother's eighty-four," I said.

"She's eighty-three, big sister."

At my brother's words, I turned toward him. Hatsuko argued that she was sure mother was six years younger than her father, who had died a week ago at eighty-eight.

"Then mother turns out to be eighty-two," I said.

"In any case, when Otane was eighteen, she was twenty-seven, so—"

"Oh, that arithmetic again."

My second son Atsushi came from Fukuoka with his wife Yasuko one afternoon when such an atmosphere still lingered, so it delighted the patient.

Using a local expression meaning "tender and lovely," mother praised her grandson's wife: "Even more *muzorashii* than the photograph." They had just married that spring, and their visit to the sickbed was also their first visit home. Otane's semiformal greeting on such an occasion, mixing Tokyo dialect and the language of ladies-in-waiting, was so fine that it could not be imitated, but when whispering to me, she returned to her native dialect.

"Otsū, it's just as your mother says. In the wedding photograph she was in a snow-white Western gown. Though she looked beautiful, there was something so lofty about her that we wouldn't dare come near her. But when we meet her, she's completely different, a really lovely person. She smiles to everyone, and what a wonderful smile. You must be happy, too, having your first daughter-in-law."

However, the attack that evening through the following day was severe. The couple found themselves in a mess totally different from the situation on their arrival the day before. Before they could tell their grandmother about their interesting visit to the family graveyard, to the main house's winery, and to the soy sauce factory on an island in the delta at the entrance to the harbor, they had to return by a late afternoon train.

"Leaving already?"

Looking weak and tired, Mother asked with an expression of surprise as the couple took their leave, with their knees together, one in his suit and the other in her dark brown afternoon dress. "He has a lecture tomorrow," I interceded.

"Of course you shouldn't skip it. So good of you to come from so far away."

Mother's eyes became moist, and two transparent drops of water sprang out. I saw her tears for the first time. Were they simply tears of contentment and regret for their leavetaking, or—? My eyes became suddenly warm as I thought about this. Trying to keep them from being seen, I left the room, hurrying the couple along.

Mother sensed the attack before it happened. Never fond of injections, she tried to endure as long as possible after the pain started, but now she asked for one even before it started. Before, she had us press her back, but now she complained about the pain in the chest. The two nurses went around to her back; others cared for her in the front. While stroking her thin chest with the ribs showing, my hand touched a nipple which was stuck there as if it were something like a rubber stop, and I was arrested by a kind of sentiment. However, even the most violent attack, once past, left her feeling normal, which was almost strange. She had, of course, no fever, and she could eat anything in the intervals. However, the faint hollows from both sides of the lips to below the cheeks began to suggest a hint of the death mask.

The weather, which often affects sick people, was changeable. When I left Tokyo, it was so mild that a thin coat felt too heavy. I was surprised by Kyushu in midwinter; then it became unseasonably warm, and then it swung back again. Within a few days, however, the sun gradually became warm again. In the yard, which had the appearance of a dry landscape garden though surrounded by black wooden walls like a town house, the buds of tall magnolias which had been in one corner since long ago turned into white blossoms branch by branch from the southern side first. It was as though white Song dynasty porcelain teacups were placed on the branches one after another.

In the guest room downstairs where one could look up at the blossoms, my younger brother, his wife, and I started a family talk about their oldest daughter Akiko. Akiko was expecting to marry in Tokyo on the thirty-first of the following month. My brother and Hatsuko were going to accompany her to Tokyo. If something that should not happen happened, a wedding was out of the question. But if mother continued to be only seriously ill, they naturally wanted their daughter to go ahead with the ceremony, if only the ceremony, with us taking their place as her parents. If that was to come to pass, I would have to return to Tokyo for preparations without losing a day. As for the patient, since she more than anyone else was pleased by her granddaughter's wedding, there was no doubt whatsoever that she would not stop me from leaving. Yet, thinking that this might become the final parting, I hesitated. Her condition improved, however, and there were even days without an attack. The fair weather continued, and her face by the screen door, through which the sun shone so brightly that it had almost a violet tint, seemed more animated than a while ago. The spring which gives life to all things might heal the disease of old age once again. As long as she caught no cold and had no problems with diarrhea there would be no immediate change, said the doctors, just as before.

Even when I finally made up my mind, I was not able to tell her the truth, so I used Fukuoka as an excuse again. I made up a story that something concerning the university, which couldn't be explained in letters, had arisen, so I was going to be away for a while. Since even when healthy mother didn't suspect people, there was no reason not to believe me when, after getting dressed up, I told her this as though I had just decided to go. She told me to thank the young people for their visit the other day, and to return quickly when I finished my business.

"Yes, yes, I'll come back right away, mother."

Told that the car had come, I put on my gloves while kneeling, looking down at mother's long, yellowish white face, with the fine nose that rose straight from the forehead in the center. Her eyebrows were growing white like her hair. She used to tell us about them: on the morning after the wedding, she shaved them by herself with a razor she had found in a drawer of the mirror stand. As for the shining black teeth, she stopped dyeing them when I was fourteen or fifteen, but she continued to shave her eyebrows meticulously. The bow-shaped attractive eyebrows reminded me of mother in my girlhood, but at the same time they shook me with a thought that this might be the last time I saw her. What was more painful than anything else was that I had to deceive her when I had come and I had to resort to something like deception again on the way home.

But at least I didn't lie about going to Fukuoka. The following day, during a leisurely breakfast of toast and red tea with Atsushi and Yasuko, I looked around in the bright sun at the house that I had come to late the night before. I praised the house as being quite good, nice and neat just as I had heard. Shun'ichi, who lectured at our son's university once or twice a year, had already seen their new home, but I had never had the chance. That was also the reason that I had to make this side trip, after having excused myself with such a fuss. Though the house was tiny with only five rooms, there was a yard surrounded by a low stone fence where vine roses would soon hang their blossoms down to the street. The peach trees in the narrow garden in back were starting to bud in soft pink, one by one, and the string beans that Yasuko told me she had planted were also already in bloom. I thought the flowers that decorated the dinner table in a round glass vase were rape, but they were turnip flowers from a corner of the same garden.

"Besides, it's almost free. I feel like taking it back to Tokyo."

The rent was so low that I made this joke. The couple also laughed and said that the flowers were all gifts left by the former tenants and that it was loveliest when the poppies bloomed.

"See, mother, those are all poppies."

Opening a screen door, Yasuko showed me outside the bright dining room. There was no fence. A little alley led to the house in back; deep green leaves with lively zigzags were growing all over on the other side of the alley and down the bank of a small stream which was the ideal image of a pastoral brook.

"They increased from the seeds that scattered, didn't they? I can imagine what a sight it must be when they turn all red. Look, there are some other things, too."

In the long narrow area enclosed with pebbles under the eaves by the side of the narrow open veranda, pansies bloomed and lilies were sprouting.

"I wish the water in the brook were clean."

"Don't ask for everything Atsushi. You're only paying nineteen yen."

We started to laugh again about this. I said that although normally a tiny house is set in undesirable surroundings, theirs was a grand place with a park with a big moat just a few minutes away by foot, and that I would love to come here to write if it were nearer. Atsushi said: It's okay if it's far; please come.

"Since I'm on the campus all day long, my room's not occupied. Free from all the interruptions you may have in Tokyo, I'm sure you'll be able to work well."

Yasuko, too, came from a scholar's family, brought up in a life-style similar to her husband's. Though living humbly without even a maid, they retained the heart to enjoy flowers and were trying to build a new life quietly, freely, and firmly. With them before my eyes, I could not help thinking afresh how different my old home was, with mother at its center.

They were living for the family rather than for themselves, living as parts of a machine that constituted the family. They were important gears for making the family work. When my father passed away twenty-five years ago, Atsushi was two: I remember his starting to toddle while we were home for the funeral. That baby now started his life like this, while another grandchild, my brother's only daughter, was soon to marry, for which I had abandoned my seriously ill mother. I saw in them lower branches and young leaves on the treetop which had started to shoot out from the trunk of an aged tree that was, in truth, decaying.

While staying another day in order to wait for a sleeping car ticket, I received a telegram from my brother telling us that the patient was doing better and better.

"I'm sure grandmother will get better again," I said to the couple, turning hope into certainty. One reason that we could enjoy a leisurely chat lay there. I reported in the same tone to my husband and Hikaru, too, who met me at Tokyo Station.

"I'm sure she'll be better again."

"But you must be tired."

"Yes, but I'm not allowed to be tired now. I've come with a big assignment."

Since I was accustomed to living withdrawn at home, it was more exhausting for me to rush out every day in the morning to get jostled in a department store crowd than to travel to Kyushu a few more times. Preparing the bride seemed to be something endless: although everything might seem ready, one missing item after another suggested itself. Having no daughters, I had never experienced that; in addition, Kase, a youth from the same area and the bridegroom to be, was graduating from college that spring, and he had to start looking for quarters, which made the fuss bigger. After many repeated trips up and down between the basement and the eighth floor of the department store, collecting items down to bean paste and salt jars, I was as done in as I would be after crossing the Japan Alps. I pleaded to the maid Osei almost pitiably:

"Osei, please take over the rest. Oh, what a chore it is to create just one bride."

Osei, who displayed the ability of the many-handed boddhisattva, went out to the apartment that Kase had rented so he could move in after the wedding and, taking a whole day, whipped their new abode into shape.

Akiko, too, came to Tokyo. On the wedding day, my brother who had once given up the idea of attending it came over, thinking that at least the father should be at the ceremony. It was also because the patient was getting better and better.

"She's much better than when you left, sister."

"I'm grateful. I'm sure she'll be all better again."

I repeated the same words.

The following day, I walked downtown accompanying my brother who, while in town, tried to buy the gifts for the couple who had acted as go-betweens. Even at one of the few best lacquer stores, they only had single pieces, unable to present a pair to us from among their expensive items. Telling us that he wouldn't know when they would be ready even if we ordered a pair custom made, for the craftsmen were at the front, the manager looked lonely with his back toward the scanty display shelves.

"It seems from now on you must think you are happy even if all you can do is just continue the ancestral trade, doesn't it."

"Our type of trade never had booms, even in the old days, so—but rumor has it some people are making profits."

"The time will come one day when such people will regret it."

Even in the shade of the spring willow on this pavement, such an exchange was no longer unbecoming. Tasting this wartime reality in the bottom of our heart, we came out of the Ginza.

My brother left on the Swallow express the following day. As we figured it, at about the time the traveler, who was to spend the night on the train, had already crossed the strait, a telegram arrived, informing us of a sudden change in the patient and asking what train my brother had taken. Crossing the reply telegram, another came, telling us of his arrival home and her critical condition. On the following day, we arranged for me to go home alone. Owing to his work, my husband would be delayed one day. I dashed to Tokyo Station, and on getting out of the taxi, I realized that I had left behind my handbag containing the ticket and my money. I left the people who had come to see me off and returned to my house in Nippori in the same car. I arrived before the gate just as a telegraph messenger was going inside, his red bike leaning against the side entrance. The result of having left behind something I rarely forget to take was that I was informed of my mother's death.

Mother had returned to the inner guest room in the main house. Though a white cloth covered her face, she lay under the same quilt as before, with a black mourning kimono spread upside-down over it. They had refrained from changing anything about her sickbed until I arrived.

"Mother, I'm home."

I greeted her with the words I was expected to say to a person still alive; my lips trembled, warm tears fell to the bedding. I lifted the cloth: her face was no different from when I had said good-bye; she looked as if she might open her eyes at any moment and say in her usual local accent, "You're back." Not suspicious of my lie that I was going as far as Fukuoka on urgent business, she had kept on waiting, repeating, "She's not back yet." Hearing about this, my tears welled up afresh. No matter what my reason was, I should not have left the sickbed then. Yet, I had left my mother with a lie, and lulled by a temporary improvement which is often deceptive, had convinced myself that she was getting better. Another part of me, however, had believed the opposite. I had wished somehow to turn my face away from this solemn fact. At the same time that I deceived mother, I had deceived myself.

Owing to the family style, even "death," instead of being painfully mourned, produced a fairly vigorous clamor.

It was not as big as my father's funeral when hundreds of individual four-legged trays were lined up for the cleansing meal, but the procedures of the funeral and the style that became a housewife were kept as in the old days. Kith and kin gathered from nearby prefectures. Since I live far away, among the wives that accompanied their young husbands there were some whom I met for the first time. As for the youth who called himself a stand-in for his father and sat politely in his school uniform, I wouldn't have known whose son he was if I hadn't heard him say his family name. What was unusual was the maids mother had employed: those still healthy or those whose whereabouts were known came over from nearby farming or fishing villages. It was customary to inform them of the death of their old mistress by dispatching a messenger to each one from the store, instead of sending a letter or a telegram. Along with the candles and incense burners, the "pillow sword"* was placed by the bedside. Mother was a Buddha all in white. After they worshiped her, with the beads in their hands, they greeted me for the first time in decades.

---

*Sword placed by the pillow for protection.

"Look who's here after such a long time. The mistress was healthy, but at last she is gone, I know how—well, don't you—?"

They grieved, wiping the tears on their old sunburned faces. The man we called Uncle Gen when we were small, husband of the woman who was the wet nurse of my cousin and sister-in-law and who also nursed me a little, joined his thick and broad hands, dyed in the sea wind, close to his familiar round nose, looking as if he wanted to pay double respects for himself and his deceased wife.

"Uncle Gen's still healthy, isn't he," I said to my brother, seeing with nostalgic surprise Gen's back in a mourning jacket whose length and sleeves were both too short and wrinkled from being stored in an old wicker suitcase. A fisherman-farmer in a nearby fishing village, he often used to bring us his catch from the sea. When five or six, I was taken with my cousin to the wet nurse's house. While watching the deep blue evening tide burst against the rocks of the white sand shore just beyond where the fishnets were being dried, I suddenly became sad, feeling as if I had come far away hundreds of miles from my mother. I remember crying, saying I wanted to go home. And, unless I remember incorrectly, it was on Uncle Gen's head that I saw a topknot for the last time as a child.

The simple funeral visits of these people made me happier than if hundreds of people of high rank and office had gathered. During the wake, former managers and close friends didn't make a move to leave, and relaxed reminiscing continued. They praised, along with her beauty and gentleness, the deceased's completion of her heaven-given eighty years, a comment which reflected the era. The adopted son of Gotō, one of the former managers who now had a big rice business, had died in action in North China. Since he was a serious youth who was praised in the neighborhood, his honorable death was all the more regretted, but Gotō said with characteristic Japanese resignation:

"Whether one lives long like the retired mistress or dies at age twenty-four like my boy, it's all a matter, I think, of predestination."

As the topic expanded, I was surprised by the many tales they had to tell about people and families in the area. For example, when the subject of another who had also died in battle came up, even third- and fourth-degree relations, let alone his immediate family background, were all detailed.

"His dad now lives in town, but he originally came from X village. You can see it from the train window; the white-walled house on the edge of the cliff above Y station is where they lived until the time of his great-grandfather who was the village's top man. In his grandfather's time, they became involved in the lime mine and lost everything. But their relatives are something. The wife of Z village's village master, I'm sure, is his aunt."

"That's right. The S town porcelain shop owner's son who's at college in Kyoto is his cousin."

"The porcelain shop owner's daughter, too, I heard, married somewhere in the Kyoto-Osaka area—"

"She married a man who works at a bank in Kobe."

The moment one of the women quickly responded to a topic, it led to a discussion, with womanly minuteness typical of such subjects, of the wedding preparations, the bridegroom's lineage, and even the virtues of an ointment that his family in the country traditionally sold and which they were said to have been taught to make by a *kappa,* river troll. No matter what they talked about, there was no malice; they simply enjoyed telling tales and had fun hearing them. Listening to their chat, which was informed by rich and precise memories, I thought that this was probably the way our ancient history and art have been orally transmitted. Among people of the same locality, however, it was not possible to conceal anything as it would have been had they lived in cities. Unlike neighborhood teams thrown together in time of war, which communicate by way of circular notices, they preserved through many generations a circular notice board without physical form.

In passing, Otane's age, which had been uncertain, became clear that evening. For, after looking into it on reporting mother's death, we came to know for sure that she was eighty-five, having been born in 1857, the fourth year of Ansei.

"If so, as Otane was eighteen when mother was twenty-seven, nine years younger, she's seventy-six."

My brother counted by folding his fingers, and told Otane, raising his voice:

"Otane, you're seventy-six, don't you forget that any more. If you forget, you're really going to lose track. For you no longer have your ruler."

The last word made everyone laugh. Otane herself, however, just grinned with her pale face. Whether she heard it or not after all the shouting, I could not tell. She looked somehow like a machine with a broken valve.

"Indeed your mother is a happy person, Otsū. She's cared for by everyone even after she's gone, you see; when I die, I think no one will cry for me."

Otane's words sounded as if right out of a puppet play. Since mother's death would indeed be a void in her life thereafter, it was understandable that she looked vacant. However, the same Otane took command with a fiercely confident rigorousness when it came to the matter of handling the body, decorating the coffin, and hosting the priests who came to read sutras at the bedside. For example, for placing the body in the coffin, three or five men, including my brother, were to wear new *yukata,* a symbol of a cleansed garment, over their clothes, with a coarse rope for a sash. The *yukata* was not to be washed right away but exposed in the evening dew overnight. White dumplings made of powdered rice and the meal offered to the Buddha had to be arrayed in a certain way, and the shiny brass flower vase had to be placed in a certain place. Otane didn't neglect a thing. Although we didn't accept a single funeral wreath, mother's ceremonial room was orderly, clean, and beautified by the decorations. What was most agreeable was the coffin. In cities, the long coffin, in order to suit the funeral car, was simply covered with white cloth, remaining bare and cold; but since the coffin was still carried on the shoulders here, it was placed in a long carriage of unfinished wood. Gold paper imitation metalwork was used as symbolic ornamentation, but the rosaries in the four corners were real. The gilded tassels flittered; a white silk curtain on which the family crest of the snake's eye was painted large was drawn around and pulled up in front with a decorative white strap. I thought mother was indeed fortunate to have lived till eighty-five, to have died leaving everyone with good feelings, and to be carried in such a carriage.

The day dawned in a dark rain. How unfortunate for the funeral guests, we thought, as we looked up at the sky after opening our eyes which had rested only in brief slumber. But it only drizzled in the afternoon. Shun'ichi arrived long before the funeral procession began. The roof of the carriage was lifted, and the coffin lid was once again opened.

"She looks as if she's sleeping."

After he honored the dead and returned to his seat saying this, I, too, went to take a last look at mother. She really looked only peacefully asleep. As I was pushing back stray wisps of her white hair, my brother on the other side of the coffin said as though asking me to do what he himself wished to do:

"Sister, stroke her face."

Brought to tears again by his tone, I quietly stroked her face as told—from the forehead to the cheek to the chin. It was stiff and cold. After all, mother was gone. I felt I realized it for the first time, but I couldn't help finding it somehow strange that her thin and tender white hair was the same as when she had lived, whether I touched it or pushed it up. The small pouch of bleached cotton hanging on her chest was a little out of place, and I took the opportunity to correct that, too. The six holed coins inside were carefully cut out of paper by Otane for the ferry across the river to the underworld. Mother had believed that every single thing down to a piece of paper and half a coin belonged to the house and was not to be privately appropriated; nor did she have any need for any of it. She never carried money for herself except when she made a trip, for example to Tokyo. If she heard a rumor about a townsman's wife somewhere secretly saving up, she scorned her lowliness. Indeed, if one searched her chest of drawers, wherever one turned things upside down, not even a cent would turn up. Not even a wallet could be found. This was the way she lived. Those paper coins were probably the only property she had ever possessed.

"Are you coming with us in this heavy rain?"

It rained so much on the morning of the ash-gathering that my younger brother came to ask me again after changing into his formal clothing. Since, according to the custom of the area, women didn't go to the shrine for the funeral the day before, but simply burned incense before the coffin left, I didn't feel like missing today's event, however hard it rained, even if we had no car.

The six of us left against the rain, we two, my younger brother, his two daughters, and Yokota, the former head brewer, who carried in a white cotton cloth a big jar which we had prepared. Though pouring down so unlike a spring rain, it was a warm and clean rain whose wetness was comforting. As we cut through town and entered a quiet street with rows of old samurai houses, cherries, peaches, apricots, and camellias were in bloom at house after house. A town with narrow slopes in an old movie called *Dr. Caligari* had reminded me of this neighborhood. Rather than a street, it was closer to a winding and bending topless tunnel on a slope, cut through a fragile lime hill. Neither the pink petals knocked down by the rain against the surface of the street of stone nor the deep red camellia whirling around in the gushing water in the little sewer were stained. I was struck by their freshness and beauty, and by that of the old-fashioned gates climbing the slope and the thick, fresh bamboo fence. Even after we approached the mountain and entered the street newly made at the foot of the mountain for the crematorium, the landscape was still attractive. There were many cherries, beautiful against evergreens' smoky light green in the now calm rain—blossoms on pines, so to speak. While appreciating the deep spring of home after many years, I pictured in my mind, tipping my indigo *janome*-umbrella with a snake-eye design, the funeral procession with mother carried in a beautiful carriage.

Mother had turned into white, light, clean bones. As I gathered them with the chopsticks, they made dry sounds like seashells. Although the glasses that had gone in together with her were not found, the silver mouthpiece of her pipe came out as a blackish lump which resembled it. I suddenly thought of taking the remains to Tokyo; after putting as much as could fit in the urn, I wrapped them in the newspaper I happened to find there. The faint warmth felt through the paper was like mother's skin not yet all cool. At the same time, I had the very odd illusion that I was holding my own bones after I was cremated.

Making tea in the guest room on returning, my brother started to talk about one thing that moved him about the funeral. The leading males in both stores, the main and the opposite stores, as well as the ones in the warehouse, had taken the initiative in "carrying their retired mistress's carriage" and everything else including the incense burner and lanterns.

"Even though it was taken for granted in father's time, we're talking about nowadays. If they said to themselves, 'who'd do such a thing,' I couldn't force them to do it. However, I was really happy since they suggested it quite unsolicited. Thanks to that, we didn't have to have mother carried by firers."

"It won't happen any longer when it's your turn."

At my joke, allowed only between brother and sister, my brother laughed a bit sadly.

"Of course that's so. Even now, it's done by firers everywhere."

I knew there lurked various contradictions that society had pressed on the family; in the fact that we did without firers, there was the difficulty that things like the relationship with employees were not as before. Though a partnership now, the company had a much more feudal interaction than that between the president and employees in a pure joint-stock company. Yet, the relationship between the family head and his servants was gone, which had allowed the head to call them by their names without honorifics while they gladly responded to that. Pointing this out to my brother, I asked him how much he was paying the men who carried the coffin today.

"No money paid." Yokota, who had joined the tea seated at the end, answered my question. "It's the rule to give them the formal clothing with the family crest we had them wear today, and let them drink a lot tonight."

They had worn the quickly prepared formal clothing of dark gray cotton with the family crest the size of a plate sewn on. Yokota continued to say that the gold brocade which was so thick as to feel heavy in the hands, at the time of the funeral of the former family head, hadn't cost four yen a piece, but that this time the paper-thin material mingled with synthetic fibers cost twice as much.

"So, how much does it cost altogether, when you do as much as you did?"

Asking this, I confided that I had guessed it might be manageable for 2,000 yen.

"That's the virture of the countryside, sister."

Laughing, my brother turned around to Ikegami, manager of the main house, who had come to say the morning greeting.

"It won't come to half as much, will it?"

"How could we let it cost as much as that?"

He answered like a merchant in a deliberately exaggerated and clownish way.

"Even including the cleansing meal of the seventh night, it won't cost that much."

"But in Tokyo you can't get such a tasty vegetarian meal with a second tray nowadays, no matter where you look."

"That doesn't cost one yen, though."

"I've decided to have my funeral here."

Everyone started to laugh at my remark, and Shun'ichi teased me saying that I must be planning to get out of my coffin to sit at the table. The vegetarian meal with the special fried dish called *hiryōzu,* sesame bean curd, *kudzu* noodles, and thick soup with malt in it—I like them so much that I would want to eat them myself at my funeral. However, even though I won't be able to eat them, and, far from being carried in that kind of carriage, my body will be carried by car and turned into ashes while people wait, still if I live to my mother's age, I'll be able to see where this rapidly changing world is going. I may be able to write something a little better, too, at my desk. Thinking of such things, I finished the now cold tea.

When it was the season to leave Tokyo for our mountain villa, I took mother's ashes. In the year when the little house was first built in the woods of a lonely plateau, mother, who happened to be in Tokyo, accompanied us and stayed there for the summer. She worried about bears coming out. That, too, became a ten-year-old story, and our summer village also changed into a humble but beautiful resort village. Mother wished to see it once again, but I purposely avoided welcoming her. In addition to a fear that she might suddenly become ill or something, I also felt lazy about keeping her so long in a totally different environment than her place. After what had happened, I blamed myself for being such an unkind daughter, and thought of at least burying her ashes in these woods. And then, after now forgetting it and now remembering it, I finally carried it out one day. In the shade of a young

Japanese oak which faced the mountains and looked down at the creek, my husband Shun'ichi and our third son Hikaru dug a hole, taking turns. I placed the urn with the ashes at the bottom.

Under
the round moon
pray you sleep.

Something like this, not precisely a haiku, came to my lips. That night the full moon was bright. Feeling that somehow we unexpectedly happened to choose a becoming day, the three of us rejoiced.

三

Blind Chinese Soldiers

Hirabayashi Taiko

On March 9, 1945, a day when by coincidence one of the biggest air raids took place, the sky over Gumma Prefecture was clear. An airplane, which might have taken off from Ota, flew along with the north wind.

Taking the road from Nashiki in the morning, I (a certain intellectual-turned-farmer) came down from Mount Akagi, where the snow in the valleys of the mountain was as hard as ice. From Kamikambara I took the Ashino-line train to Kiryu, transferred to the Ryoge-line, and got off at Takasaki. I was to transfer again to the Shingo-line to go to Ueno.

It was around four-thirty in the afternoon. Although the sky was still so light as to appear white, the dusty roofs of this machinery-producing town and the spaces among the leaves of the evergreen trees were getting dark. The waiting room on the platform was dark and crowded with people who had large bamboo trunks or packages of vegetables on their shoulders or beside them on the floor. It reverberated with noise and commotion.

After taking a look at the large clock hanging in front of me, I was about to leave the waiting room. Just at that moment, a group of policemen with straps around their chins crossed a bridge of the station and came down to the platform. Among them were the police chief and his subordinate, carrying iron helmets on their backs and wearing white gloves. The subordinate was talking about something with the station

---

Mō chūgoku hie (1946). Translated by Noriko Mizuta Lippit with the permission of Shinko Teshirogi. The translation first appeared in the *Bulletin of Concerned Asian Scholars,* Vol. 12, No. 4 (1980).

clerk who accompanied them, but it seemed that the word of the police chief, who interrupted their talk, decided the matter. The clerk crossed the bridge and then returned from the office with a piece of white chalk in his hand. Pushing people aside, he started drawing a white line on the platform.

I was standing in front of the stairway with one leg bent; I had sprained it when someone dropped a bag of nails in the crowded Ryoge-line train. The clerk came up to me, pushed me back aggressively, and drew a white line. As was usual in those days, the train was delayed considerably. The passengers, quite used to the arrogance of the clerks, stepped aside without much resistance and, to pass the time, watched what was happening with curiosity.

Shortly, a dirty, snow-topped train arrived. Before I noticed it, the policemen, who had been gathered together in a black mass, separated into two groups. They stood at the two entrances of the car that I was planning to board. The white lines had been drawn right there.

The car seemed quite empty, but when I tried to enter I found myself forcibly prevented by a policeman. I then realized that in the center of the car there was a young, gentle-looking officer sitting and facing another young officer who was obviously his attendant. With his characteristic nose, he was immediately recognizable as Prince Takamatsu.

With the strange, deep emotion that one might experience upon recognizing an existence hitherto believed to be fictitious, I gazed at this beautiful young man. My natural urge was to shout and tell everyone out loud, "The Prince is in there. He's real!" Yet it was not the time, either for myself or for the other passengers, for such an outburst. Unless one managed to get into one of the cars—at the risk of life and limb—one would have to wait additional long hours; how long, no one knew.

I rushed to one of the middle cars immediately. Yet my motion was slowed by the wasteful mental vacuum that the shock of seeing the Prince had created. I stood at the very end of the line of passengers, looking into the center of the car and trying to see whether there was some way I could get in.

After glimpsing the pleasant and elegant atmosphere in the well-cleaned car with blue cushions, I found myself reacting with a particularly strong feeling of disgust to the dirtiness and confusion of this car. Shattered window glass, the door with a rough board nailed to it instead of glass, a crying child, an old woman sitting on her baggage, a chest of drawers wrapped in a large *furoshiki* cloth, an unwrapped broom—a military policeman appeared, shouting that there was still more space left in the middle of the car, but no one responded to his urging.

I gave up trying to get into this car and ran to the last car. There were no passengers standing there. A soldier, possibly a lower-ranking officer, was counting with slight movements of his head the number of the plain soldiers in white clothes who were coming out of the car. An unbearable smell arose from the line of the soldiers who, carrying blankets across their shoulders, had layers of filth on their skin—filth which one could easily have scraped off.

I was looking up at the doorway wondering what this could mean; then my legs began trembling with horror and disgust.

Looking at them carefully, I could see that all of these soldiers were blind; each one stretched a trembling hand forward to touch the back of the soldier ahead. They looked extremely tired and pale; from their blinking eyes tears were falling and their hair had grown long. It was hard to tell how old they were, but I thought they must be between thirty-five and fifty years old. On further examination, I observed that there was one normal person for every five blind men. The normal ones wore military uniforms which, although of the same color as Japanese uniforms, were slightly different from them. They held sticks in their hands.

Judging from the way they scolded the blind soldiers or watched how the line was moving, I guessed they must be caretakers or managers of the blind soldiers.

"Kuai kuaide! Kuai kuaide!" [Quickly, quickly] a soldier with a stick shouted, poking the soldier in front of him. I realized then that all the soldiers in this group were Chinese. I understood why, even aside from the feeling evoked by their extreme dirtiness, they looked strange and different.

All the soldiers who were led out of the car were left standing on the platform. There were about five hundred of them. I doubted my own eyes and looked at them again carefully. All of them half-closed their eyes as if it were too bright, and tears were dripping from every eye. It was certain that every one of them was blind.

The supervising soldiers who were not blind saluted suddenly, and a Japanese officer with a saber at his waist appeared from one of the cars.

"What about the others?" he asked, passing by a soldier who was busy counting the number of blind soldiers.

"They will come later, sir, on such and such a train," the lower officer answered.

"What on earth is this all about?" the sympathetic yet suspicious expressions of the passengers seemed to ask. A middle-aged woman even started crying, holding her hand-towel to her eyes. It was obvious that both the commander and the lower officer wanted to hide the blind soldiers from the passengers, but it took a long time to get the rest off the train, and the number of onlookers gathering behind the fence gradually increased.

At last those at the head of the line began climbing up the stairs of the station, while the train started moving slowly. I was standing on the steps of the car in front of the one which had just been emptied and was holding on with all my might. I could see the policemen who were guarding the soldiers whispering to each other.

"I guess they were used for a poison gas experiment or they are the victims of some sort of explosion," said a man with an iron helmet on his back, standing four or five persons ahead of me.

"They don't have to carry out poison gas experiments in the motherland," a man who appeared to be his companion objected. Following up the companion's comment, I asked a woman of about forty who was standing next to me,

"When did those soldiers get on the train?"

"Let's see, I think at around Shinonoi."

"Then they must have come from around the Nagoya area," I said to myself, although it did not give me a clue to understand anything.

Soon the passengers forgot about it and began to converse.

"I came from Echigo. I am on my way to Chiba with my daughter." The woman whom I had just come to know started talking in a friendly manner. She told me that she was bringing her daughter to report for duty in the women's volunteer army and that her departure had been delayed for a week because her daughter had had an ugly growth on her neck. Since they could not get through tickets to Chiba, they would go as far as they could, then stay in the place they had reached, standing in line until they could buy tickets to continue their journey. They had come this far, she said, but the hardships they had been through were beyond description.

I had been offended a moment ago by the unconcerned way in which this woman had answered my question about the Chinese soldiers, but I now thought I could understand it. The Japanese were too involved in their own affairs to be moved by such an incident.

When the train left a station some time later, I went into the car which had been occupied by the Chinese soldiers, hoping to sit down and rest. I returned soon, however, because the smell there was intolerable.

The conductor came from the end of the train, announcing "Jimbobara next, Jimbobara next," as he passed among the passengers. By that time, the windows on the west side were burning with the rays of the setting sun, and the huge red sun was setting with the sanctity of the apocalypse. I realized that the car occupied by the Chinese had been taken away and that my car had become the last of the train.

Yes, there was the Prince, still in the car ahead of us, I remembered. But I was too tired to tell anyone.

After the war was over, I asked the merchants who had their shops in front of the Takasaki station whether they had seen the group of Chinese soldiers boarding the train again. They all said they had never seen them again. Perhaps they never returned from that place.

Narcissus

Hayashi Fumiko

Since he just stood with a cigarette in his mouth by Tamae's bedside, she had to ask, lifing her eyes from the magazine, "How'd it go?" Sakuo was not so tender as to report: Mother, it was thus and so today. So she asked, "How'd it go?" but he grinned, dragging a cushion with the tip of his toe toward him. "No good, because you're a parent with no paternal piety, honey." He stuck out his red tongue and kept on exhaling rings of smoke.

"What do you mean 'with no paternal piety'? How can you call your mama honey?—Did you meet Mr. Tsuda?" "Of course I did. He says hi to you, honey." Tamae rose and remained gazing at Sakuo's face for a while. "What's 'without paternal piety'?" "There's an expression in this society, 'without filial piety'—It's the opposite of it. Mama here, I'm saying, has no paternal piety—"

Tamae began to boil inside at his spiteful way of talking, despite the fact that he was her own son.

"I have no reason to be called by such a name. Papa left us early, so didn't I bring you up to this day with my female hands alone? You pick on me as nobody else does, just like papa did. You're already twenty-two. Why don't you try to get a job and settle down for my sake? It always ends in failure because your attitude's wrong. As Mr. Kamiyama said, 'Come to ask for a job and smoke a cigarette looking cocky like that—that's enough to spoil everything—' "

"I see, then, job hunting is just a matter of attitude, is it? Are you saying that any fool'll do if his attitude's all right?" Sakuo had pro-

---

Suisen (1949). Translated by Kyoko Iriye Selden with the permission of Ryokubin Hayashi.

vocatively long sideburns, a haircut like a Spanish matador's. Soft hair grew under his nose in such a way that it was hard to say he was completely grown. It was hard for Tamae to think he had been a child going to elementary school just a while ago. It started to feel as if another man from somewhere else was sitting by her bedside. "Why is it that you can't behave with humility? Nobody thinks well of you." "That's because you brought me up this way, so it can't be helped. You've raised an ideal son; I bet you have no complaints, have you?" Her eyes full of tears, Tamae watched her son who sat with his legs crossed. Leaning against the wall, he was almost hatefully calm. "A parent and a child, we are just the two of us. Why do we always have to argue like this?—What did Mr. Tsuda say?" "Nothing. Just that he couldn't do anything for me so I should try the test; so I took the test." "Do you have confidence?" "None. They asked such stupid questions that I didn't feel like writing anything—" With a somewhat weak grin, Sakuo poked what was left of his cigarette into the dirty ashtray. "I wish you were good at exams at least—" Her son's dull face made Tamae think wistfully that he would never be successful in his long life. "If your attitude's bad, and your exams are also poor, even Mr. Tsuda won't be able to do a thing, will he?" "Well, that's about right—" Tamae no longer felt like talking. A long time ago, she had felt like abandoning this child; if she had boldly given him away to someone then, she wouldn't be in this much trouble now. "What kind of questions did you get?" "What kind? Things I have no interest in, like how will France solve her crisis, or how old is Truman. I've never even thought about such things." Feeling bitter, Tamae stood up, reeling, wearing over her wrinkled towel nightwear a heavy silk kimono jacket, also soiled. She went to the washroom to put some water in the kettle.

When she returned, Sakuo was rummaging inside her handbag. "What are you looking for?" "I need a little money." "There's no money no matter how you look. You really have become a bad boy. Please stop torturing your mother. Why do you want to pick on me so much? What do you mean?" Finding a nail file in the handbag, Sakuo wildly rubbed his nails with it. "You're joking, aren't you?— Nobody's picking on mama. I promised to go to Sakurai's tonight, and I thought a grown man like me can't take just the train fare—" Tamae no longer answered. As she put the kettle on the electric heater, she stood in front of the mirror hanging on a peg, staring at herself without moving her eyes. Old age was sneaking up on her. Tamae somehow regretted her age; she was forty-three. She felt as if she had just aged idly. It was not that she had aged while frantically working for her child's sake; but if only she didn't have this child, who knows but that she might not be leading a happy life? At this late stage, no matter how frantically she might struggle, surely a woman's happiness was never to come again.

Taking a comb to her rough, lusterless hair, she stroked it down. Her hairline had become markedly thin. Reflecting that it might be because of her intemperate conduct when young, Tamae applied hair oil and tried to get her bangs to hang fully over her forehead. Her lean face with pointy cheekbones seemed somewhat younger. Then she combed the bangs back, trying to give herself a changed look with a surgeon's boldness. Strangely she suddenly aged. The water in the kettle boiled, so she poured hot water into a basin, steamed a towel, and covered her face with it. Under the hot towel, her eyelids twitched. "Are you going out somewhere, Mama?" "Yeah, I'm going out money-making." "Have you got any prospects?" "No, but I can't help it—" Removing the towel, Tamae looked at herself in the mirror. Her face had perked up, ruddiness coming into her skin. It would be nice if her skin always looked like this; but she pasted kerosene-smelling cold cream thickly all over it. Her knuckled fingers repeatedly massaged around the rim of the eyes on her face shining with oil. Again letting her bangs hang loosely down when she had finally finished making up her face, she dotted her wrinkled eyelids with rouge and stepped back to look at herself in the mirror. The murkily dirty lukewarm water in the basin, suggesting poverty, revolted her. Perhaps because it was already five days since she'd gone to the bathhouse, her made-up face didn't look as fresh as expected. The lipstick did not spread at all on her rough and hardened lips, so, the nether lip protruding, they looked like discolored raw tuna slices.

"Mama's no longer so strong, though you may think I'm dependable. Wouldn't you rather leave me and rent a room at Sakurai's or somewhere? Mama's really tired. You and mama were born to meet as enemies; it was fated from a former life. Don't you want to liberate me now that you're a grown man? No matter what you do, I won't complain, and as for me, I too don't want you to complain no matter what I do. The only thing I worry about is that you aren't very healthy. But when you get sick, that's another matter; I'll deal with it somehow. Listen, won't you somehow live independently? I think that will be better for you, too—" "You've started to feel I'm in your way, have you?—" Sakuo said. "There's no special need to separate; besides, if you want, I'll call you 'big sister' like I used to—" Tamae, applying lipstick thickly in desperation, looked at her lips in the mirror, revealing her white teeth. She gave no answer. Sakuo said, "I don't want to leave—"

Looking at her face in the mirror, her eyes shining with malice,

Tamae said, "However, mama no longer has the patience to stay in the same room with you. I'm not young enough to live while constantly fighting, and besides, like in Maupassant's *Woman's Life* which you also read, I feel as if I'll be killed by you." "What a joke. You're so conceited. Who'd kill his own mother? I don't think of you any more than I do of the dirt in my nails—I just borrowed your stomach, like Christ—" "O yeah? I see. Then get some wings, feel free to fly away from your Mary. I have no reason to feed you now that you're grown up." Tamae wore a fairly worn-out green jacket and black slacks. As she sat in front of the electric heater, her feet in lightweight woolen socks, she daubed her nails red, holding her hands over the heated electric coil. Sakuo watched her hands abstractedly. It was hard to think of them as anything but the devil's hands. The memory of having never received a gentle word from this mother's mouth stimulated antagonism: he was filled with a desire to torment her thoroughly. "What dirty hands—you're aging." "Mind your own business. Who made my hands so dirty? These are the hands which go to work, come back, handle firewood, and feed you." "I see— Is that why they get so dirty?" Having finished manicuring, Tamae rubbed her shiny, enameled nails with a rag and again put both hands in the lukewarm water. She tried to wash her hands thoroughly with a cake of soap which yielded few suds. Then she wiped her hands with a towel and held them away from her eyes to scrutinize them.

"Say, I'm not kidding. Go wherever you like, please. Really, I'm sick and tired of you." Sakuo, his eyes closed, kept lightly knocking his head against the wall for a while. Tamae threw her coat over her shoulder, poured the remaining water from the kettle into a cup, and drank it with rationed brown sugar. "You've never even once thought of me with affection, have you, Mama?" She suddenly let the cup fall from her lips. Tamae looked at her son. He looked very tired. He looked something like her divorced husband in his youth. "Of course, mama has at times loved you. But when I'm chased around by life, there're times when I can't be engrossed just in you. Of course, since you're the child who hurt my tummy, I think fondly of you, Saku. But the time has come when we have to part. Mama thinks so— You were mama's Saku only when you were a small boy. You too now have a grown-up's eye, so you look at me as though you were a stranger— Mama and Saku, don't you see, we're this kind of parent and son. I have no desire to end where I am; I have plenty of desire to work. When I face your maliciousness, I come to no good. Besides, you're a burden to me, Saku—" Taking a cigarette case from the pocket of his tattered overcoat, Sakuo offered it to her. Tamae took a cigarette and held it in her mouth. Sakuo, too, held one in his mouth and struck a match to light both. "All right. Then I have no choice but to go to Eiko's; is that okay?" Hearing him mention Eiko, Tamae answered, "Yeah. You have no other way? Then do it. When your mother can't feed you, you'll have Eiko feed you? How come you can't work like a grown-up instead of always depending on a woman? Even if Eiko is infatuated with you, you can't trust that woman, and for another thing there's so much age difference between you two— When people ask me about her, I feel embarassed."

Tamae was born in Taibei, Taiwan. Her father worked for the railroad, and she received a rigorous upbringing as a government official's daughter. She became acquainted with Ibe Naoki who, after graduating from a theological school in Amoy, was staying a month or so at the home of someone she knew and the two came to Tokyo, eloping after a fashion. This was the spring when Tamae was nineteen, a year after she had graduated from girls' school. They rented a modest house for fifteen yen a month in Oimatsu, Zōshigaya. Tamae bore Sakuo there. Ibe wished to go to the United States, but too poor to do so, he barely supported his family on his slender income from helping to edit a theological magazine. Tamae's school friend Ōkawa Tazuko came to Tokyo to study for the entrance examination for a music school. While she stayed on the second floor of Tamae's house, Ibe and Tazuko formed a relationship which they kept concealed from Tamae. After two months or so, Tamae discovered their relationship and berated the gentle Tazuko every day. Tazuko was a passive woman. After she began having relations with Ibe, she wilted day after day, and soon lost all desire to enter music school, succumbing to indolence. Apparently unable to live in the same house, Tazuko moved out, finding a small house with three rooms or so in Dōzaka, Hongō. It was Dōzaka, but near the stone buddha of Tabata, where many workers commuted to downtown wholesale stores. She moved while Tamae was out, but before long Tamae scented out the Dōzaka house and there was a scene: she stormed into the house and pulled Tazuko around by the hair. Within half a year after moving, Tazuko committed suicide with gas in the house. Ibe quit his job at once, left for Amoy like a whiff of wind, went from Amoy to Malaya, and then on to Kuala Lumpur, rumor had it. It was around 1928. Tamae hated Ibe, and forever hated Tazuko who had died. Whenever she recalled Tazuko's whitish face with freckles around the eyes and nose, Tamae was haunted by the thought of the deceased who, she heard, had a rubber gas pipe in her mouth. Seeing light freckles near Sakuo's eyes, Tamae felt repelled, thinking it was as if Tazuko's resentment was clinging to him. Tamae helped edit a religious magazine for an acquaintance of Ibe, but it was difficult to make ends meet for her son and herself plus a maid. Since she had broken off with her family in Taibei, she earned her living precariously as though walking a tightrope from man to man.

When Sakuo turned from a child into a boy, gaining understanding, his mother's life was inexplicable in his eyes. From the perspective of his boylike puritanism, it was uncomfortable. Tamae had the middle-school boy call her "big sister." When they started to live by themselves, no longer keeping the maid, Sakuo cursed his mother who frequently slept elsewhere. The two of them lived together for a long time railing at one another. Having been disappointed at home, Sakuo was too timid to mingle with people like his schoolmates. He tended to neglect academic work, and even after just barely graduating from middle school, he lived lazily, depending on Tamae and enrolling in a private school called B Academy. During the war, since he was weak, he wasn't drafted for labor service. Evading laws of that sort, he moved from place to place with Tamae.

After the war, Tamae rented a room in an acquaintance's house in Kōenji and lived with Sakuo. She found a job in the paper, something like head maid in a love hotel in Ikebukuro, and commuted from home. She approached black marketeers who came to the hotel, and started to deal in foreign drugs. She gradually saved some money, but whenever she saved enough to give a sigh of relief, Sakuo spent it all. The relationship between the two grew more ominous day by day, and sometimes Tamae thought she wanted to kill him. She didn't know if Ibe was alive or dead; she hadn't heard from him in over twenty years.

Sakuo had a woman lately. He unabashedly told Tamae that she was a dancer with a husband. One day when she chanced to return home early and found a woman sleeping with Sakuo, she grabbed the hair of her son's woman who was asleep and railed at her harshly, just as she had at Tazuko long ago. The woman never came again, frightened by Tamae's fierceness, but Sakuo told Tamae, as though to make her uncomfortable, that she was threatening to leave her husband to live with Sakuo. Sakuo didn't particularly like her; he had no feeling of love. He simply chose her as a convenient woman who fed him. Having been brought up by a wild mother, he didn't know real love. Lazy like an old man from an early age, he waited in all matters till a woman took the initiative.

"If you want to separate, it's fine with me. But it can't be done today, all at once. For one thing, I have to go to Eiko's to consult, so it's not simple." Tamae started to calculate; once she parted with Sakuo, everything would be the way she wanted it. There would be no need to pay a lot for a room to have a tryst with Tomita. With the money, she wanted to redecorate her room so that it would be a little more pleasant to live in. The dust she swept out from the room to the porch to the narrow yard had piled up in a heap. A stray cat often came and poked at it. The fence of Japanese cypress around the narrow yard had been left to wither. Once a sneak thief came in through a slit in the fence and ran off with her only pair of shoes. Since she was so sloppy, she was not allowed to use the landlord's kitchen. She had been bedridden for the last few days. Perhaps from age, she just stayed in bed, in no mood to do anything. The New Year was approaching, but there was no feel of it in her place. She didn't even think about celebrating with a piece of rice cake.

Though she got ready, meaning to go out, she was intimidated about going downtown wearing clogs on the muddy streets after yesterday's rain. "Don't you really feel like getting a job, Saku?" Sakuo was whistling quietly, shaking his shoulders. "I don't feel like it. Everything revolts me. It's a drag even to keep on living like this. But after all, I feel lonesome unless either mama or Eiko is hovering around me. Even someone to fight with is better than no one," he chuckled. "If you feel like black marketing, I'll ask Mr. Tomita about it. How about doing something like that?" Sakuo continued shaking his shoulders and knees for a while, but said, "I've tried selling fountain pens in Shibuya with Sakurai. Sakurai's big brother found the job. They didn't sell at all. It was ridiculous. I have no talent for business. When I asked Eiko she sold as many as six. After all, it sells better when a nice-looking woman does it." Tamae was amused by the way Sakuo, apparently thinking his woman attractive, had no qualms about speaking well of her. "O yeah? Is Eiko nice-looking? I wonder what's so lovely about such a fatty. Isn't she like swollen with water?—" "To me she appears a beauty, so it's okay. Rice cake skin, smooth belly, beautiful skin." "You think so because you knew a woman for the first time." "But even when you were young, you weren't as beautiful as she—" Tamae opened the sliding porch doors, numb from the absurdity of it all. The sky was clear. It was so warm that vapor seemed to rise and dance above the dark soil of the farmland. Red underwear that Tamae had hung under the eaves twisted like a persimmon peel. Probably because it had been there for days, it looked a bit dirty.

In the late afternoon, Tamae and Sakuo left together. She went to Kichijōji in the opposite direction from where Sakuo was headed. She took the wide avenue and then cut through the park to Yuki's house. She had the superintendent call Tomita at once, but the message was that he had gone away on business a few days earlier and would not return until around the third of January. Thinking it odd, since they had planned to meet at Yuki's house once more before the New Year, she called the Kayaba-machi office herself. A woman answered and said that Mr. Tomita was out with a cold for four or five days. She didn't know which to believe, but she had an uneasy feeling that something was wrong. She went near Tomita's house in Shimorenjaku and stood watching by the wall at the corner for people coming out of the gate. A young woman in slacks came out. Pretending to be a messenger from the office, she said, "I have a message from the company for Mr. Tomita. Is he in?" "Oh, no, the master is away on a business trip. He is expected home around the third of January—" "Well, that's too bad. I was sent on an urgent errand from the company. Then it wasn't a business trip for the company after all?" "You may be right. How odd. I'll ask the mistress—" The maid went back toward the gate. Tamae quickly turned around and hastened toward the station on the road along the dark narrow canal, which smelled of water.

When she got off the train at Kōenji, she changed her mind. She suddenly thought of dropping by at Kamiyama's. He was having supper, looking glossy after a bath. Drinking sake with some herring roe, he offered a cup to Tamae, too, saying, "How about a drink?" Kamiyama's wife looked cool around the neck with her hair up, having just been to the beauty parlor for a permanent. "No matter when I come, it's cheerful here. I envy you, Atsuko, you're happy." Atsuko, Kamiyama's wife, had gone to the same girls' school in Taibei, and they had become close because she and Tamae both lived in Kōenji. Kamiyama worked for an economics magazine in Marunouchi and had tried to get Sakuo a job, but shocked by his unparalleled indelicacy, he never offered to help him again. Learning that their classmate Tazuko had been driven to suicide by gas by Tamae, Atsuko kept a certain distance from this strong-willed person, but she could not just coldly watch her friend fall apart before her eyes. Atsuko resembled Kōgō, the Empress, so she was called "Kōgō-san" by her family members. Tamae accepted the cup of sake, and Kamiyama refilled it several times. "It's still too early for you to age, Tamae, but why didn't you remarry earlier?" Kamiyama said. "Because I have a child." "Who cares if you have a child?" Atsuko said. "No, my Saku is very evil. He always watches me and spoils anything like that." "How foolish can he get?—It's natural for a child to wish for the mother's happiness," Kamiyama said. "But mine's different. He's queer." Tamae didn't feel like going home to a dark cold house. While in bed imagining, she dreamed of brave deeds, but once back in reality, nothing interested her. Each time she saw a sprightly young woman walking cheerfully with a man, she felt jealousy as if toward a woman of her age. No matter what, she no longer had that kind of youth. Why she aged and sank so quickly, she could not explain even to herself. But she raged as if Sakuo had ruined her life. Human beings have thought of various things, but after all, everyone only plods along the road of greed. Sitting at the Kamiyamas' cheerful table, she hoped she might be invited to eat with them. If she ate there, all she would have to do was go home and sleep.

In any case, that night she badly missed Tomita now that she found she couldn't see him. Perhaps it was all over between them. Once, Tomita said, "Your body has become very bony." If she tightly pinched the flabby skin of her arm, it did not spring back quickly like it used to, remaining wrinkled like tired rubber. Though she thought of training her body somehow, she was too wrapped up in her livelihood, and she spent day after day lazily doing nothing but arguing with Sakuo.

It was around nine, after supper, when she left the Kamiyamas'. Perhaps because she had imposed, the couple didn't even see her to the entrance. She walked on the wide road that ran through the cedars where the houses came to an end. The cold wind, chilled by the frost, rose, and countless fine stars were shining in the sky. A tall man followed her. Tamae embraced a certain hope. She imagined she might be spoken to. She walked slowly, appearing deep in thought. When they came to the post office, the man who was following her passed, looked back at her face by the light of the street lamp, and then went away. He was young. Tamae somehow felt cheated.

On returning home, she went in through the back and opened a glass door. The electric heater was shining in the dark like a one-eyed monster. Sakuo seemed to have crawled into bed. "Saku?" "Yeah." "How come you're back?" "Eiko says 'not today.' " When she switched on the light, Sakuo's eyes looked swollen as if from crying. "How dangerous. It's better to put a kettle or something on the heater." "Have you got a cigarette?" "Not one. You're in no position to smoke when you don't even work." Since it was cold, Tamae, too, threw her coat over the thin covers of her bed and crawled in. "On the way back I went to Sakurai's. He asked me if I felt like going to Hokkaido. There's a job in a mine office. They have a workers' dormitory, and the pay's good too, he says." "Well it sounds attractive, doesn't it? Why don't you go? Do they ask about your schooling?" "I gather I can just write something plausible—" "It's good for you to go to that kind of place and train your body. Where in Hokkaido?" "Bihoro." "I see. It's going to get colder from now on; I wonder if it's hard for someone who has a poor chest." "I bet mama'll be happy if I go to that kind of place. Good riddance—" "Right, right. Whatever you think's right, of course—Your mama's a horrible woman; she'll feel relieved when you're gone." Tamae was getting irritated from hearing the noise of the mahjong tiles inside. "Going to Bihoro is like going to die—" "Not at all. It may even be better than living in Tokyo. If there's something good, please call me. Mama's really tired of Tokyo." A fire engine passed by beyond the vegetable field, whirring and shaking the ground. "Hey, Saku, dear." "What?" "Mama's very lonely. Though you won't understand this loneliness—I felt so disgusted today. I'm so strong-willed that I've had lots of bad luck in my life. Somehow, I've grown tired of living. I'm quickly getting ugly, and I don't have the spirit I once had. You're a man, so you understand men's feelings, but men are pitiless." "Now you've realized it?" "Yeah, now I've realized it. Is it just when they're young, between man and woman? I wonder if it's just that kind of thing. Papa, for one, went where he wanted to go, abandoning you. He was easily fascinated with a woman and was irresponsible. I feel so sad about human beings." "All you need is money. God made human beings invent this ingenious thing. Every problem can be solved if only you have money. Eiko, too, said she would leave her husband right this moment if only she had money." Tamae opened her fingers with manicured nails like a fan, and gazed at them. She looked as much as to say that she could not show her dirty and wrinkled hands to a man.

"It's hard to tell whether Saku is in fact a good or a bad man." "I'm bad." "Not necessarily. You're still only twenty-two and haven't had much experience in bad dealings; but can't you trick a millionaire's daughter or somebody?" "Ugh. I hate daughters." "That's because you've never had a young woman." "Mama, you're a scoundrel—" "It seems." Tamae felt she didn't care what bad things she might do now. In ten years, she would lose the spirit for it. She felt that everyone was misled by a hypocritical morality. Amid hypocrisy, human beings fought like furious lions for the bait called control, power, and wealth. That vitality of human beings almost steams with joy and harmony. There is laughter in it. However, there was not a single hope for Tamae and her son. Not even in the parent-child relationship—

"Do you feel like going?" Sakuo didn't answer. He watched the ceiling in silence. Tamae wondered till what age she used to sleep with him. She had never touched his skin since he was around six. He always slept alone quietly. He knew neither ordinary happiness nor common civilized manners. She had never taken him on a vacation trip. Yet he was precocious, and she knew that he showed signs of enjoying himself alone from age sixteen or seventeen. If a human being had nothing else to enjoy, probably it was natural, Tamae thought, that he fell into that kind of thing, and she pretended not to notice. She did not ask where he had eaten that night.

A couple of days later, Sakuo really prepared to go to Hokkaido. He brought home 3,000 yen from Eiko, something like solatium. "When are you leaving?" "On the thirtieth, by the night train. With Sakurai. I won't return." "Yeah, let's keep well—" For the first time tears filled her eyes. It was not that she wanted him to stay, but she was moved, despite herself, and was filled with tears of instinct.

On the night of his departure, they walked in the crowded year-end Ginza. "Women are terribly pretty." "There are many beauties in Hokkaido, too." "City women are nice after all. They'll all have men, I guess—" Tamae and Sakuo talked like friends, walking shoulder to shoulder. "It'll already be cold with the snow, I bet." "Well, I can't tell till I get there—but I can warm myself by the coal stove; isn't that luxurious." "Born in Taiwan, I know no colder place than Tokyo, but a place in deep snow is a little romantic, isn't it—" "Romantic—not so, I guess, if I live there— What are you going to do from now on?" "I'll age more and more. It won't be like before. I may suffer from occasional fits of asthma during the long winter and then die suddenly." Sakuo bought two packs of Hikari cigarettes and put one in Tamae's hand. "You won't be able to have a man any more." "You're right. No more hope, nothing promising can be expected. How's your time?" Sakuo looked through the window at the clock in a clothes shop. "Still okay, I've got at least two hours." "I won't see you off, all right?" "Yeah, that's better. Eiko's coming to see me off, so that's better." "Did you say good-bye to Eiko?" "Yeah, I went there this morning. She's supposed to wait at Colombin's there." Tamae stopped in her tracks. She didn't feel like seeing Eiko. Suddenly various feelings raced through her mind.

"Say, if anything happens to mama, you needn't come back— You know me, there may be a moment when I feel like dying in a fit of passion. But you don't have to come, Saku." Sakuo nodded with his chin. Wearing a dusty, worm-eaten beret, he looked wretched to her though he was her own son. It was perhaps a blessing that there was a woman, even though another's wife, on his departure from Tokyo, Tamae thought, and took his hand. Sakuo shook hands with her rather weakly and released her hand right away. "I won't see you for a while, but stay well. I'm not a good correspondent, so I won't write," she said. On a tree-lined avenue in front of a coffee shop smelling of coffee, Sakuo quickly disappeared into the dark, carrying his things wrapped in a *furoshiki*. Tamae looked back once or twice, but soon lost sight of him in the night fog.

—She was completely alone, she thought, and raising her shoulders, she let out a deep breath. Even back streets were crowded at the end of the year, and a store with silver-colored salmons hung under a blue light, a display of mannequins with black velvet over both arms, and what not streamed vacantly past her eyes. Downtown at year-end seemed not to have changed at all from the old days. Tamae for no special reason thought of herself breathing her last somewhere in this wild city. Only that kind of thing would, it seemed, express her final youth. In her mind's eye, it was associated with a candle blown out by the wind. In the December city street, a girl wearing a traditional hairdo and a couple of children were noisily playing battledore and shuttlecock. In the light of the lamp under the eaves, the white-feathered shuttlecock now shone and now disappeared. —Coming out on Fourth Avenue, she heard a saleswoman straining her voice in front of Morinaga's which was black with a throng: "Here are the old familiar Morinaga Velvets. Wouldn't you like some candy?" Tamae, slipping into the crowd, picked up a shiny cellophaned pack and put it in her pocket. She felt very good. At a chinaware store, she stole a cute Kutani soy sauce pourer, unnoticed in the crowd. Not because nobody caught her, but because of the weight in her pocket, she felt good. She felt as though she was walking wearing a mask. Somehow she even started to feel it was pleasant to live. Liberated by separating from her son, at the same time, she felt as if she had suddenly become young. As she came to a dim Sukiyabashi street, Tamae took a velvet from the cellophaned pack and put it in her mouth. The sweet melody of a popular song came from the lighted advertisements. The *Asahi*'s neon news ran busily to the right, flashing the dissolution of the Diet in the sky.

五

Residues of Squalor

Ōta Yōko

Around the little ramshackle shed, there was a splashing sound of rain. It looked like a torrent.

At midnight, when the world was hushed in sleep, this awful-looking hovel did not stand alone in solitude, for around it hundreds of identical houses stood side by side, and human beings were alive, asleep in those houses. When I thought of this, my mind was, unusually, at peace. Speaking of solitude, the whole cluster of hundreds of houses beaten by the June rain was, I should say, enwrapped in solitude. I felt as though the warm breaths of those asleep reached my skin from neighboring houses not one iota different from the one in which I was sitting. Maybe at least one person was awake in each family, I thought, and like me was engrossed in killing the slugs that were creeping around in the shabby, rain-soaked house.

I peered into the adjacent room. Since I had brought the only electric bulb in the house to the three-mat dining room next to the kitchen, there was only dim light on part of the old pale green mosquito net hung from corner to corner filling the adjacent six-mat room. Between the lighted spot and the shadowy rim, I could see countless slugs creeping. They clung to the skirt of the aged mosquito net and slowly climbed the surface of the net, creeping as slugs do. Filing up at a fixed interval, they silently crept one after another, sticking all over the mosquito net, their molluscan bodies slowly undulating. Every bit of mois-

---

Zanshū tenten (1954). Translated by Kyoko Iriye Selden. This story has benefited from the translation assistance of Rebecca Jennison.

ture seemed to serve as the slugs' food and air.

Inside the mosquito net, five relatives were sleeping together, almost piled on one another. From the side of the dining room where I sat, closest to me was the face of my younger sister Teiko, asleep. Her face was close to the net, her nose pressing against it. She was a young widow with two-year-old and six-year-old girls. The two girls were sleeping in a disorderly manner, half crouched, near their mother's feet. Perhaps to avoid the electric light, my mother lay facing away from the others so that my eyes only caught sight of an old *kaimaki,* a kimono-like comforter with sleeves, covering her legs. The other sleeper was a female houseguest who was not ordinarily here, my younger cousin Hashimoto Miyano, who had come from the country that morning to see me on my return from Tokyo after three years. Not yet having been able to tell me all about the changes in her fate, she seemed to be asleep with her back toward me on the far edge of the space inside the mosquito net crawling with slugs. There seemed no space for me to sleep inside the net. Miyano lay where I usually slept. She seemed to scrunch up her body intending to leave some room for me, and there was a slight space between her and the edge of the mosquito net.

A bunch of blood relations were asleep scrunched up under one mosquito net, each shouldering a misfortune that was unthinkable in normal situations; on that net a group of slugs, also unthinkable in normal times, was creeping around. Needless to say, I was aghast at this sight. However, I objected to the method my mother and my sister Teiko used, annoyed by the infestation of the slugs: they prepared salt water in an empty can and, picking them up with throw-away wooden chopsticks, dropped them in. I did not want to kill slugs. I wanted somehow to save their lives. Slugs knew nothing; they were innocent. After the defeat in the war, Japanese typhoons with female foreign names came and devastated various places year after year. In the ruins of H city, which had suffered a unique war disaster, too, torrential rains poured, and typhoons blew. On the ruins of the old parade ground in the city of death which had been reduced to nothing, houses for those whose homes were burnt down were quickly built, leaving the rubble of the destroyed army buildings which had once stood close together. Not everyone camping out could get into those shacks. Teiko and her husband barely managed to get into one by winning the black market lottery.

Houses for bombed-out people were built all over one corner of the spacious former parade ground. Exposed to heavy rain and stormy winds each year during the rainy season, every house crumbled and rotted while slugs bred under the floor. Since this was a valley, there was no drain for the water. The floorboards, steeped in mud throughout the year, were starting to rot. Slugs were breeding there in swarms. My mother and Teiko diligently dropped them in a can of salt water. I looked in the can. They were half melting, but not completely melted. Thick and muddy, there was no sign of their having put up resistance to this sole primitive measure. After once eyeing this sight, I had begun to suffer from an association about human beings heaped up in a mound of death, half burnt but not completely melted, with no energy to show any sign of resistance. They were so alike. I could not think of slugs as mere slugs.

Heartless mollusks to me seemed personifications with hearts, and I could not bring myself to kill them by sprinkling them with salt. I made sure that the people in the mosquito net were asleep and no one saw what I was doing. I had bought DDT during the day. If I repeatedly and carefully sprinkled DDT all over the room instead of massacring them with salt, which must be a horrible shock to slugs, eventually they would give up entering the room, I thought. When I mentioned DDT, Teiko and my mother were silent. It was palpably clear that, cornered into penurious living, they thought salt was less expensive than DDT. Having touched that chilling reality, sadness weighed heavily on my heart, but I had to try it while they were asleep. Peeling the seal of the cylindrical DDT container, I shot white powder onto the dining room floor. Turning and twisting, slugs had encroached in groups from the side of the door sill under the glass doors—there were no wooden rain shutters. The door sills on the four sides of the room were not even well grooved. They crept all over the dining room, climbed pillars, and clung to the legs of the low dining table. Their traces shone, forming many glittering streaks. Since the slugs were mostly swarming in the slits between the glass doors and the door sill that led directly to the outside, I repeatedly sprayed white power there. The slugs would run away, I thought. I immediately realized my ignorance. While the strong smell of the chemical attacked my eyes and nose, the soft bodies of the slugs on which it was sprinkled slowly melted and flowed.

They could not leap or fly like fleas or mosquitoes; nor were their boneless bodies able to resist the stimulus of the chemical. I felt nauseous. This was what I feared most. On the day the pale white radioactive flash burnt H city as though to toast it, I was in that city, and I saw how human beings were burnt and melted not by flames of fire but by the rays of the homicidal weapon that had fallen from the sky. I had been suffering from the intense shock for six years. It was now June 1951; but for me every moment of the long hours after the war was dark. Trying to escape from that dark affliction, I sometimes took sleeping pills, even during the day, and gave myself injections of antihistamines which were narcotic. I tried drinking but found that it affected the stomach before I could get drunk enough to efface the pain.

I dreamed of escape in death. That very spring, a poet* who had experienced the same thing, and who seemed to have been afflicted by the same pain, killed himself. Instantly, I thought I had fallen one step behind. Now that he had died, I thought, I, also an author, could not follow suit. When I went to his wake, I wanted to question him about the meaning of his death, but the poet who had killed himself was no longer there. Realizing that death meant that he was gone forever, at that moment I thought suicide ugly.

I stopped sprinkling DDT and turned my eyes away from the melting slugs. Still I could no longer tolerate it. Folding a newspaper narrowly, I covered that place. I had to forget the presence of the slugs as quickly as possible. It was because I recalled the groups of humans massacred seven years ago. On the night of my return from Tokyo about ten days ago, my mother bought a little sake for me. I wanted to drink it in one gulp and paralyze some portion of my nerves. I peered toward the mosquito net. Teiko was looking at me, her eyes wide open inside the net.

"What's the time?" Teiko asked. She didn't seem to have noticed that I had sprinkled DDT and melted slugs.

"Twelve-thirty."

"Somehow the smell of a chemical woke me. I wonder what it is."

"I just sprinkled DDT around because there were fleas and mosquitoes. Say, why don't you get up for a minute? Let's drink," I said, without even a smile. I wanted quickly to finish turning the corner of my heart that was filled with dark depression.

"Shall we?" Looking faintly despairing as though in response to my face which suggested despair, Teiko, who hardly ever drank sake, lifted the skirt of the net and came out. She went to the kitchen and started to make a fire in a pitch black earthen hibachi.

"You needn't heat the sake. Let's drink it cold."

"No need to stand on ceremony. Making a fire's nothing."

"It's not ceremony. Sit down quickly."

---

*Hara Tamiki (1905–1951), Hiroshima-born poet and author, was in Hiroshima in 1945, thinking, as did Ōta, that it was safer than in Tokyo. Hara's "Summer Flowers" (Natsu no hana, 1947), translated by George Saito with additions by the editor, and "The Land of Heart's Desire" (Shingan no kuni, 1951), translated by John Bester, are available in *Atomic Aftermath, Short Stories about Hiroshima and Nagasaki*, ed. Ōe Kenzaburo (Shueisha, 1984).

With sake in a glass bottle and sake cups, Teiko sat at the small dining table facing me. She stretched her arm behind her and took out damp peanuts from the cupboard. Teiko and I lifted our cups of cold sake and, after letting them touch with a click, drank silently.

"Everything's cold, isn't it?" I said.

"These peanuts are too damp. Wait a second." Teiko rose lightheartedly, sliced a cucumber in random shapes, dotted it with salt, and brought it over. The green peel of the cucumber looked refreshing.

"Raining, isn't it?"

"Yes, it rains so much."

"If it rains too badly, the mud'll be washed away and human bones will come out, don't you think?"

"They come out even if it doesn't rain."

"Even after decades, if someone digs in the ruins of the parade ground, human bones'll still come out, won't they?"

"You don't have to wait decades. Even now, when we dig around here to make a vegetable garden, lots of things like *bijoh* come out besides bones. Slates and eating utensils, too."

I didn't immediately grasp the meaning of *bijoh.*

"The buckle for soldiers' *obikawa* leather belts. They called *obikawa 'taikaku,'* giving it a Japanized Chinese reading. Military terms are all disgusting, aren't they?"

"Lots of *bijoh* of *taikaku* come out? What I'm saying is that not now but decades from now, lots of them will still be unearthed—*bijoh* and eating utensils."

"Even after hundreds of years, I'm sure human bones and soldiers' *bijoh* and eating utensils will come out."

These words conveyed the numbers of soldiers and other humans who died inside and outside the military buildings that were in this parade ground. Tetsuji, younger brother to me and older brother to Teiko, met instant death on August 6 six years ago at the site of the First Unit, the present housing site for those who were bombed out, without even having his bones identified. As we drank cold sake, however, neither Teiko nor I said a word about this. We did not want to wake the wailing that had sunk to the bottom of our hearts. Even among blood relatives, people refrained from talking about their grief and shed unbearable tears, perhaps in bed, when the world was quietly asleep. Sōichi, Teiko's husband, had been called at one point during the war to the military hospital that was on this parade ground. It was adjacent to the First Unit where Tetsuji died instantly. Since Sōichi was in the army in Kyushu on August 6, the day of the atomic bomb, he did not die in combat. Soon after the war he suddenly died of tuberculosis. Sometimes I connected his death with war, and I had the illusion that his was a kind of death in action. While entertaining this illusion, I thought it was no illusion, and the deaths of Tetsuji and Sōichi were imprinted in my heart, as two deaths in combat, heterogeneous yet overlapping.

"Oh no, I'm getting itchy," Teiko said, her face red from not even two cups of sake, suddenly starting to scratch a corner of her lip with her right hand. Right at the edge of her mouth a deep cut forming an x-shaped keloid mark pulled in an ugly manner. At the time of the bomb, she was cut in more than thirty places all over her body by glass splinters which flew at her like knives. From the center of the x-shaped cut in the lip, water and medicine that she put into her mouth poured out. My mother and I had said it was good that Teiko was already married. However, now that Sōichi was dead, what we said invoked the contrary psychological response in us.

"It's bad to drink sake, isn't it? Oh, it itches."

"Shall I give you a shot for hives?" I meant the antihistamine which I habitually used to sleep.

"I don't want a shot. Must you give yourself those shots? Poisonous to the health, no?"

"Control poison with poison. I can't live straight, can I?"

I was a little drunk. The procession of slugs creeping around came vaguely into sight. Teiko went out to the kitchen and drank gulps of water.

"Does it still itch?"

"It's a little better now."

Teiko saw that I was somewhat more cheerful. As though waiting for a chance to tell me about it, she suddenly started in a soft tone, "Mr. Kurata, you know."

"Mr. Kurata?" I thought. It was no longer possible that my heart leaped at that name. Yet the name seemed to penetrate gently into my chest.

"Huh?" I said to my sister, who was more than ten years younger. I was expectant.

"About a month before you came this time, Big Sister Mitsue saw him at the streetcar stop in Misasa."

"So he's alive."

There was a lie in this. I had heard it rumored that he had survived August 6.

"Yes, she said that he was dressed up as much as before. He knew that you went back to Tokyo after the war and are writing, but wondered if you were fine. He asked the same thing twice. Then he also asked about mother, and said he felt sorry for mother about you, now that his children are getting big, especially his daughter being in the prime of maidenhood, the time for her to marry drawing near—I hear he was on the verge of tears. Big Sister Mitsue said she felt sorry."

My younger sister Mitsue, older than Teiko, who married into the Misasa Shrine, was now a mother of five children. From her childhood till about the time she married, she had seen enough of my relationship with Kurata. He was a man who, in a sense, messed up my destiny. I did not forgive him for hiding the fact that he had a wife and children in the country and for machinating, so to speak, to carry on a bigamous affair with me when I was only twenty. Without forgiving him I loved him, and bore him a child. Now I would leave his house and then again be taken back. I spent eight years of my youth in this way. When we parted for the last time, my parents did not consent to my bringing home the child. My mother hated Kurata. She tried to make me hate him, too. Kurata hated my mother, who played a big role in our separation. He gave our child to a couple who were close friends. The couple feared my seeing the child. Although I loved the child even after the separation, I continued to love Kurata even more. To forget him, I took to whiskey and absinthe, which were unfamiliar to me, and while destroying part of my integrity, I tried to change myself through writing. My writing was not enhanced. I tried not to love a man. Literature that would move oneself and others could not exist where there was no love for fellow humans.

A long time passed, and in the midst of the Second World War my child should have been seventeen or eighteen. Unless I saw him then, I knew I might miss him for good, for he was at a dangerous age when he might die as a boy kamikaze pilot. I wrote a letter to the child's foster parents. A reply came. It said that the child had no knowledge of Kurata's and my past, and that even now he thought his foster parents his real parents. If I was to see him, they wanted me to see him unbeknownst to him as a complete stranger. That would be the saddest way of seeing him, and I hated the idea. I gave up the notion of getting together with him. Several years earlier, Kurata's wife, whom he had divorced once but remarried after his separation from me, had died of illness. Kurata's oldest son and the younger sister of Kurata's deceased wife started to visit me often.

Despite the fact that one of them must have been aware of his mother's resentment toward me and the other of her sister's, the two young people occasionally brought me news about Kurata and my child, as though talking about their relatives. Kurata married a young wife. Kurata's first son repeated that this woman resembled me.

Separated by months and years, the intense memory of love from long ago had faded in my maturity. Only nostalgia remained toward the fact that my youth once had such dynamic moments.

When Teiko mentioned Kurata's name and told me what he had said, I only felt as I would toward a close relative, my thought having changed to unruffled reminiscence.

"He asked her to send heartfelt greetings to mother. He said he wished to be forgiven about the old days, Mitsue said."

If so, he would no longer chase me. My heart sank somewhat, and I felt compelled to grab at a void, realizing this was the outcome of that love.

"I want to hear about the child."

I thought the child I gave birth to might be dead. Perhaps they refused to let me see him because they did not want me to learn about his death—I sensed the shadow of this possibility. If he had died, I should not try to see him. I felt that part of my heart had been supported solely by the awareness that a child who was mine was alive on this globe.

"Mr. Kurata seems to live near the Misasa Shrine. Mitsue said she might run into him again. Shall I tell Mitsue to try asking him about your child?" Teiko said simple-mindedly.

"I wonder if he'd tell the truth if Mitsue asked. Tsuruko's younger sister Naoko came again before I left Tokyo. I asked Naoko, too, but . . . "

Tsuruko was Kurata's wife, who had been tormented by Kurata and me until she died. Her sister Naoko said that Kurata didn't know my child's whereabouts after the war. She seemed to know that the aged foster father who lived in Kobe worked in the proofreading section of an Osaka press, but beyond that she kept her mouth closed. I told Teiko the name of my present companion, who did not necessarily always live with me:

"He says he wants to be there too when I see my child. But if the boy's dead, then I'll be more shaken up about learning it than by not seeing him, won't I?"

Suppose he had died at the front, or suppose he had happened to come to H city from Kobe on some errand and met his inevitable final moment due to the atomic bomb. Since this was not impossible, I was afraid of pursuing only to deepen the wound. With the war disaster further complicating life's problems, my soul was utterly confused.

Having drunk most of the sake, I became lighthearted, and since I was prone to laughing when drunk, I suddenly laughed, shifting my mood.

"Let's wake Miyano. I wonder what would be best for her to do from now on. No matter where we turn, none of our relatives is doing well. Wake her, wake her," I said to Teiko. We heard Miyano's voice from inside the mosquito net:

"I was just thinking of rising. You two were mumbling and mumbling. I have been awake for a long time."

Miyano, cold at first glance, always spoke politely like a stranger, although she and I were cousins of about the same age. She crawled out from the net with a somehow despairing pale face, nearly identical to Teiko's and mine.

"Won't you have some sake? We've left a few cups."

"Sake? I am afraid sake is . . . "

She sat upright by the table. Single-minded, absolutely serious, which was her temperament, she looked as much as to say that it bothered her even to look at something like sake. Whenever I saw Miyano I recalled something. On August 3, six years ago, she and her mother, my mother's older sister, stayed overnight at my mother's and Teiko's house. I had also returned there from Tokyo, where air raids were fierce. After sleeping on the spacious second floor, both left H city early on the morning of the fourth. They had managed to send their furniture to the country in two horse-drawn carts. Having vacated their house, they were escaping from the city. Wearing work trousers and carrying rucksacks, the two trodded off. My seventy-year-old aunt had tied her *setta,* strong leather-lined sandals, to her feet. As we thought this was the last time we would see them, a faint pain nagged at everyone's heart. After the war, this aunt lost the family's entire fortune, which she had placed in an account with the South Manchurian Railway, and died of illness soon after discovering the loss.

"The sick man is waiting at home. After visiting Tsuyako at the hospital with you and shopping a little, I have to go home as soon as possible. So I want to ask your opinion tonight."

I found Miyano's tragic situation after the war ironic. From childhood, she lived in Fengtian, Manchuria, her father and older brother occupying important posts at the South Manchurian Railway. At nineteen, Miyano married a local wholesale drug dealer. However, when it was discovered that he was dealing in illegal opium, Miyano's father, who was an extremely religious Buddhist, hated the deception and took her back to his house. After that Miyano remained unmarried for a long time. Her father and brother both died, her father from old age in Fengtian and her brother from tuberculosis at the Japan Red Cross Hospital of H city. The Manchurian Incident showed signs of enlarging the war, and, chased by the flames of war, Miyano and her mother returned to H city. They lived in H city for a long time. Since they had many relatives, they recommended that Miyano remarry, one by one bringing concrete proposals for candidates. Miyano refused to marry, and at times she declined so angrily that the bearer of the offer withdrew, shocked. She sharply rejected marriage as *kichanamashii,* which meant impure in our dialect, inviting laughter from all relatives, who commented that marriage was not simply a matter of impurity. When a man wanted to marry her, she said he must be staking out her father's inheritance left with the South Manchurian Railway. Miyano opened a sewing class and took students. Having contracted vertebral caries, she wore a cast and a corset. Around the time the war ended, the infection stopped, improving her complexion and making her plump. She was nearly forty then.

A man lived alone near their house in the country. He was a bachelor and middle landlord who had never married because of the tuberculosis he had suffered in Tokyo in his youth, and he was engaged in humble many-sided farming. He was not yet fifty. Drawn to Miyano during the war, he brought her grains, vegetables, and eggs. Before dying at an old age, her mother suggested that Miyano marry this kind and lonesome man. Bereft of her mother, when Miyano realized that after all she could not draw even a cent of the savings she and her mother had left in a single South Manchurian Railway account, for the first time she seemed to realize her isolation. She consulted as many relatives as possible by writing or talking. I received a letter in Tokyo. I replied affirmatively. What a nice mate she had found at this late stage, I thought, vulgarly envying the luck of a man and a woman of mature age marrying almost as if for the first time. After marrying, Miyano never wrote of her feelings about the marriage. At the end of the second New Year's card after moving into that house, she added that her husband had had a stroke, though light, and was bedridden, half his body paralyzed. I had been concerned about the possible relapse of tuberculosis which both of them had suffered. I had not expected to hear, however, that in a little over a year after their marriage half her husband's body would become paralyzed. The news of unpredictable human affairs left my heart overcast. Asking someone to care for the sick man, Miyano had rushed to see me, running with small steps, so to speak.

"Just because he got sick, you can't very well abandon him, can you?" I had no choice but to say.

"I have no intention of abandoning him, nor do I have a home to return to, but there is no telling whether he will lie in bed for five or ten years, is there? Suppose he stays like this as long as ten years, and I care for him? When he dies, they may tell me, you have no children, so please leave. Then what?"

"Would anyone say such a thing?"

"In that kind of situation, there is always some relative who might. He has many relatives. You write novels, don't you? Please tell me if you have a good idea."

"Even if I write novels, I can't see everything," I laughed a little. I thought that Miyano perhaps found it worse than meaningless, perhaps found it a grave loss, that her husband had collapsed after one-and-a-half year's marriage.

"Even while caring for the invalid, I have to farm; otherwise we can't eat. I feel miserable because it is as if I had lost myself again, all the way at the end."

Miyano did not love her husband, I thought. Thinness of affection was reflected here and there in her words. I said a little cruelly, "But you can't tell who'll die first, can you?"

"That is so. At our age."

Teiko silently listened to our brief conversation. She was by nature reticent. Never commenting on Miyano's broken life, she crawled stealthily under the mosquito net, looking sleepy. As though following Teiko's back with her eyes, Miyano turned her face to the net.

"Oh," she raised her voice, surprised. This was because she saw slugs not just around us in the dining room but climbing up all over the mosquito net and gathering near the ceiling.

"How awful. I never saw so many slugs in my life. I am getting a shiver," she said, narrowing her shoulders.

"It's good that you were not in H city on August 6, Miyano. When looking at this swarm of slugs, I tremble thinking that soldiers who died on this parade ground are back in the shape of slugs."

It was still drizzling the following day. Miyano and I went out into the city in the rain to visit our younger cousin Tsuyako who had been hospitalized for three full years in the Red Cross Hospital. Having decided to buy some food for Tsuyako at a store on a busy street, we walked along the streetcar road. A dingy little streetcar tottered like a lazy man in the middle of the Fifty Meter Road now under construction. A strange landscape was in view. In the center of the city where the traces of devastation were still raw, a wide road called by that name was being constructed with no apparent purpose. The miserable roadside stands that stood helter-skelter in the postwar confusion had already been moved to the riverbank, which flooded every year during the rainy season. Only a strange-looking Chinese-style noodle shop remained there all by itself.

Around the middle of the not yet finished Fifty Meter Road, close to the streetcar rails, the dirt was raised just there, and the little Chinese-style noodle shop stood tilted on top of the mound. I had learned from a reporter, with whom I walked on this road four or five days ago, why this shop alone remained there. It was not that the shop was left unheeded. The others disappeared on receiving a small amount of money for being forced to move, but the noodle shop sat there, refusing to accept compensation. The man who ran the shop was from a third country. Referring to the present freedom of the third-country person, the reporter laughed gleefully as though he was the victor. The house atop the dirt forming a tiny hill had dirt steps in front and back, down to the Fifty Meter Road. On the streetcar side of the tilted and almost collapsing shed, a signboard for Chinese noodles hung. A German shepherd, leashed on top of the steps carved on the back side, was lying flat. Since the dog was out of proportion in comparison with the shed, it looked larger than the shop. Rhododendrons and garden jalaps were in bloom from around the shed to the steps. When I briefly explained about the shop, Miyano, who was walking at my side, laughed.

"Isn't it interesting that only the third-country person is holding out like that."

"But do you know why just that house is standing on the raised dirt mound?"

"Why?"

"It's said that the whole city rose three feet. In other words, the Fifty Meter Road is at the original level. That house still stands on the raised place, but underneath—" I closed my mouth.

"Bodies? Shambles?"

"Both."

We passed by, suddenly averting our eyes from the Chinese-style noodle shop atop the mound of dirt. The rain did not stop falling. Men who looked like construction workers were planting saplings in the rain along the Fifty Meter Road for a future tree-lined street. At first I did not understand that they were to line the street. I asked an old worker as we passed them.

"What are you planting?"

"Saplings to line the street."

A shadow of pathos stole into my heart. Although the constant sorrow never left me, at this moment it shook my heart more strongly than ever. The fact that new trees had to be planted along the new road, now after seven years, overlapped in my mind with the character of this Fifty Meter Road. Like others, I could not overcome the premonition that the road might be for future military purposes.

"What trees are they?" I again asked the man planting them in the rain.

"Plane and linden. We won't live until these become big trees and grow thick foliage, though."

The old man looked up at my face. I, too, looked back at his half-crying, half-laughing tan face.

At the entrance to the Japan Red Cross Hospital we asked at information where Tsuyako's room was. I found it surprising that Miyano had never visited this woman during the three years of Tsuyako's hospitalization. I was angry myself that none of us relatives sufficiently visited Tsuyako, who had tuberculosis of the kidneys and whose throat was also now affected.

Yet I had not stopped by at her room when I recently visited this hospital. With a girl from this city whose face had lost its shape due to the atomic bomb, I came to visit another girl whose appearance had also become half-human and who was having an operation at this hospital at this late date. On that day I had not felt like visiting Tsuyako. Since Tsuyako, the daughter of my mother's youngest brother, was born in Seoul, I had never met her. From photographs I had occasionally seen, I knew that she was a beautiful woman who resembled her mother. However, my feeling toward her was not necessarily unadulterated. Twenty years ago when, having left Kurata with much agony, I started to write a little in Tokyo, my father, who was a landlord, lost everything and died. Money sent to me in Tokyo became scanty. I fell ill. There was no prospect of my novels selling. I was trying to live by writing. I wrote my greedy uncle, who had bought land and several houses in colonial Seoul, asking for money. In spite of the fact that I knew he was greedy from my own childhood experience, I had dreams about a cousin who lived far away. No reply came. I wrote repeating the same words. My uncle's wife answered. She wrote that they could not respond if a niece who had never even sent them midsummer and year-end cards asked to borrow money.

When Tsuyako reached marriageable age, my uncle and aunt, who were then in Sariwon, Hwanghae-do, and did not wish their daughter to marry in a colony, wrote asking me to find someone for her in Tokyo. I owned a house and lived with my mother then. I asked my mother to write to my uncle that, if he cared for his daughter, he should have cared for another's daughter, too. My mother wrote nothing of the kind. If I thought of my uncle in this way, my mother said comforting me, I should take care of my young cousin. Carrying a much too formal photograph of Tsuyako in a long *furisode*-sleeved kimono, my mother and I made the rounds of the houses of relatives and acquaintances looking for someone for Tsuyako to marry. A relative's family became interested. Letters and photographs were speedily exchanged between

Tokyo and Korea, and a young man was chosen. He became definite. He and his mother were about to go to Korea to see Tsuyako. Then suddenly my uncle wrote that Tsuyako was preparing to marry a promising youth who worked for the South Manchurian Railway.

After the war, my uncle and aunt were among the first to return by black-market boat. They had lost everything. They were both past sixty. Now they were both teachers in an elementary school branch on a plateau in the mountains of H Prefecture, where there wasn't even electricity. Of Tsuyako's two children, they kept the six-year-old. I no longer resented those people who would not even send five yen as a token of sympathy to me who had asked for one hundred yen. However, a trace of the scar of life that had been carved in my heart twenty years ago still remained, pulled taut. With my pity for Tsuyako nuanced, I felt somewhat concerned about our meeting.

On opening the door to what was called a model ward, we saw, in a certain kind of fresh cleanliness maintained in the large, bright room, five gloomy, pale, and lean women lying quietly. Both Miyano and I spotted Tsuyako immediately. It was not possible to hide mutual recognition of the resemblance between something in her features and ours.

When Miyano and I entered the room, Tsuyako was crouching, having gotten down on the far side of the bed. She seemed to recognize us immediately when our eyes met, but, smiling faintly with a fleeting expression, she remained crouched. Tsuyako was not as drawn as I had imagined. After finishing passing water, she politely greeted me on her bed on our first meeting and exchanged words as an old acquaintance with Miyano. Since I was no good at greeting anyone properly, I was flustered by this and quickly asked about her illness. Tsuyako answered, now in bed.

"I have to pass water thirty-five or thirty-six times every twenty-four hours, so I don't have time to sleep."

"How old were you when you first took ill?" I asked my younger cousin, who had just passed thirty.

"About a year after my parents left Korea, I also left with my two children and went home high in the mountains where they live. After another year or so, my legs were incredibly cold, and I began to pass water frequently," Tsuyako said in a frank and honest manner.

"My parents live on a plateau deep in the mountains where things are inconvenient, you see. There was no doctor, and it was a job to visit a doctor in town. By the time I saw a doctor, it was fairly bad. After staying in the doctor's clinic as a patient about half a year, I was hospitalized here."

"I wonder why you contracted the kind of illness that affected both your kidneys," I said full of compassion.

"—Before crossing the 38th parallel, my husband died of tuberculosis."

"I heard about it from my mother."

"After that I crossed the 38th parallel with two small children, with or without food, and I had a really hard time getting to Pusan. That kind of thing, I feel, was also a cause of the illness," Tsuyako said calmly, as if talking of someone else.

"I have sent my boy to my husband's parents' place, but I feel sorry for my father because I'm in this shape. If I recover, I would love to live with my children in the mountains where my parents are, perhaps as an elementary school teacher. If I don't get better, I will be a burden on my father until I die," Tsuyako said, as though telling the thoughts that were on her mind all this while.

Tsuyako also slowly spoke, intermittently, about her father, who worked at a branch school deep in the mountains. He got up at three in the morning whenever he came to visit her at the hospital. He walked twenty miles down the mountain trail in rubber-soled workers' *tabi,* changing his clothes and shoes before boarding a bus. Miyano and I said to Tsuyako that she shouldn't talk too much now. Despite the fact that I pitied her, I did not feel my chest so gnawed by this individual's misfortune. At heart, I involuntarily compared Tsuyako with the young woman who was still groaning in a ward almost straight above this one, whose entire body had been burnt by atomic bomb radiation. When I first saw maidens who appeared half-human due to radiation, I felt as though I had been pierced through and cried despite the presence of onlookers, thinking of their present and future. I rarely cried about my younger brother Tetsuji as an individual. When compelled to out of resentment for the sake of all, I cried alone at midnight.

While looking at Tsuyako and listening to her story, I could not help noticing that here, too, the war disaster pierced the abdominal walls of life in general. For me, Tsuyako was in one corner of my resentment of the whole. She was certainly included in the circle of that resentment. There lurked in me a certain psychology that was difficult to tell others. It was difficult to tell others, since it related to death. A poet who witnessed the same thing in this city had reawakened this city's devilish memories from the beginning of the Korean War, and he had killed himself as though unable to withstand the traumatic sense of crisis he harbored about the future. Around the same time, a similar psychology developed inside me. I disliked suicide. However, the thought that I might commit suicide or that I might suddenly die due to the resurgence of the atomic bomb disease never left my mind. There was a black dot in my heart concerning this return to H city. It meant a valediction to my home, ruined city though it was—my ultimate silent valediction to my close relatives and dear acquaintances. I wished to see them once again without making it look final.

"Is there anything we can do?" I asked about Tsuyako's immediate needs, but she answered with a light shake of the head, "Thank you. But I am fine because the nurses take care of everything in this model ward."

When we said good-bye and went out to the corridor outside the ward, I thought it was good to have seen her, although I was also aware that my visit took a conscious effort.

"The moment I glanced at her, I realized that she looked exactly like you," I whispered to Miyano as we walked along the corridor toward the entrance.

"You look more like her," Miyano instantly replied. "You have the same face."

We went out to the spacious entrance which was now restored to its prebomb appearance. As we were going down the low stone steps, a white-robed nurse whose face I remembered well passed in front of the flower bed. She was one of the head nurses who had served me food and told me of many things when I came last. Since I remembered her name, I thought of calling to her. I didn't. It was because her words heard on my last visit darkly flapped their wings in my heart, and I could not bear to hear them repeated. The young nurse, named Fukuhara, had shown me the traces of the deep wounds on her limbs incurred under the debris. Lying beneath the wreckage of the nurses' dormitory and encircled by fire, it was not on August 6 but on the evening of the 7th that she had crawled out to the hospital entrance.

"Even doctors, nurses, and patients who survived were half-dead and bloody. We didn't know what to do. Wounded outpatients kept coming in droves. Since I couldn't stand up and walk, I tended them by crawling around. Every single one said please give me water, let me drink water. Water, water, they said, as they died one by one. Like watering plants, I crawled around pouring water into their mouths saying, Here, water, here, open your mouth. Here, water. Here, water, But finally there was no longer a drop of water anywhere. I hate it! I shouted in a loud voice and picked up a baby from the side of the body of a mother. What were you born for? Weren't you in the warm stomach for nine months and finally born? You shouldn't have been born, I said to the baby."

Fukuhara's feelings directly touched my heart.

"The dead filled the corridor, each with a piece of paper that said 'Dead.' " A crazy man started to hop over them. Referring to that flash he said, 'A wide yellow sash wound around him,' and went hopping over the bodies as if they were stepping stones, hugging his head. The following day he said, 'Lemme have so-u-up,' and died while having it."

There was no limit to Fukuhara's stories about the misery at the hospital.

If I called to her and if she resumed her story from that time, I felt that it would burden Miyano, who was to return to the country before the day was over, with cruel images. Before accompanying her shopping, I called Shunkichi from a department store pay phone. Shunkichi was Kurata's son who visited me often, sometimes staying overnight, after Tsuruko's death. My mother, who by then had forgiven Kurata, was kind to Shunkichi. He worked at a small company, founded after the war in H city, as *senmu,* executive director, a title which had a postwar ring. He answered the phone as if he had been waiting for the call.

"I saw your name in a newspaper panel the other day, so I knew you were back. I didn't know where you were, though. I wrote you a card care of the newspaper," he said lightly like the wind. Arrogant as ever, I thought. I said I wanted to see him once.

"It doesn't have to be once. Won't you come to my place today? In the evening fireflies come out and it's lovely," he said. I knew Shunkichi's temperament because I had lived with him and the other children during my marriage to Kurata: his words were mixed with diplomatic flattery. Shunkichi said he would come and get me where I was. The more he insisted, the more I half doubted his true intentions. Part of me pondered learning of my child's whereabouts by seeing him and easing the nostalgia toward my old love, which still lingered in me. I said to Shunkichi on the other end of the line:

"I'm going to be with a relative for two hours or so. She's returning to the country, so, after seeing her off, time permitting, I'll come to your place."

In a lively way Shunkichi gave me directions to his house in the suburbs. Miyano was shopping inside the department store. Soon she and I went out to the entrance and opened the umbrellas we used both for rain and sun, Miyano her rusty red one and I my green one. We went to the rainy streetcar road and got on a streetcar. I don't know when I'll see this cousin again, I thought. Then I began wishing to see Kurata. It was not the choking passion of old. It was a flow of sweet feeling with a tinge of intimacy like that I harbored toward a close relative. I thought that Kurata might understand my dark state of mind as a result of the war disasters.

Kurata, a dozen years older than I and a follower of socialist ideology in the line of Ōsugi Sakae, an early twentieth century anarchist, first mentioned the socialist's name when I was leaving him to go to Tokyo for the first time, and he suggested that I try to see him. Although Kurata dropped out of every movement due to his naturally nihilistic character, I wanted to believe that the throbbing of antiwar sentiment was still burning in his soul.

Shunkichi had a family, a wife and children, and lived apart from Kurata, but I thought I might somehow chance to encounter Kurata at Shunkichi's.

We arrived at the depot of the bus that would take Miyano back to the country. Even after sitting down on the waiting room bench I remained silent, but realizing that we didn't have much time, I said in a small voice:

"In your case, you never know what happens when. Living alone with an invalid must be strange, but, well, I guess you should stay with him till he dies."

"Yes, well, I guess that's so."

"There's no escaping. If you escape, you'll only regret it."

In a sense I also applied these words to myself. At four in the afternoon, the bus Miyano got on left. I recalled Shunkichi's words that he would be home by four. If I were to go to Shunkichi's house at the foot of a mountain range on the other side of the big river which flowed from a mountain village, I would have to take a streetcar and then transfer to a bus. To get back to Teiko's house, it wasn't much of a walk to the site of the old campground in the grass after getting off the streetcar. Returning to Teiko's house where slugs were creeping and visiting Shunkichi's house were both depressing. Hesitantly I bought a gift for Shunkichi's children.

Rain was also drizzling at the foot of the mountain near twilight. Three bicycles stood side by side at the entrance to Shunkichi's house. One was a beautiful bicycle with red enamel paint on the body. One was a child's bicycle. With a Go board placed in the center of the guest room without proper furniture, and with a Go handbook by his side, Shunkichi was alone placing black and white stones. Under the eaves on the verandah side hung a cage with a canary. Stones and moss were placed deliberately in a small yard. I perceived an odd atmosphere.

"Somehow this is oddly old-man-like," I said to Shunkichi, who was putting away the Go board.

"Yeah, I've aged suddenly."

"How old are you?"

"I'm thirty-two. Girl students call me *ojisan.*"

He was the age of Kurata when I came to know Kurata.

"At thirty-two you're into this kind of old people's life-style? It's as if you're almost sixty."

"That's how old my old man is."

"At thirty-two your old man wasn't arranging Go stones alone as if he'd been abandoned. He despised caring for bonsai plants."

"I've aged early because I don't know when I'll die."

As he said this with a fleeting smile, I saw the appearance of the weak-minded Shunkichi, who resembled the beautiful but not striking Tsuruko and took after Kurata's tall, angular body and big, bony hands and feet. Although it annoyed me that his face resembled his mother Tsuruko's and his bodily features his father Kurata's, I was also surprised to see how enervated he was, sitting cross-legged in front of my eyes.

"My hair fell out completely once, but while I was anticipating death, hair after hair started to grow, you see, and I've managed to live seven years."

In this city, no matter who meets whom, this kind of dialogue cannot be avoided.

"I stayed in this house the whole time. But my old man was in the mountains of Ujina. Do you remember that August 6 is his birthday? My wife made *ohagi* rice dumplings for him early in the morning and left, saying that she was going to bring them to Ujina—that was before eight o'clock. The train seems to have gone through the area that became the target unexpectedly fast, and by 8:15 when the bomb fell it was fairly close to Ujina. But I didn't know it, so I thought she must have died near the target area. From the following day I walked around searching for her from Hatchobori to Kamiyacho, which were still on fire. The dead, the dead, there wasn't even space to walk. I couldn't tell men from women, so I looked into each face."

Shunkichi was exposed to the residual radiation in the target area, he said.

"On looking back, because the whole city burned for three days, and even after the flames subsided the situation was such that women could not walk, I understand that she couldn't walk home from Ujina. Since she didn't come home for three days, I thought she was dead and looked for her body for three days in a row."

"Your wife was in Ujina?"

"Yeah, she came back looking innocent with the old man on the fourth day. But I began to spit blood after a month."

"When I think of that radiation disease, I can never forgive America," I said, my voice trembling.

"Radiation disease is something like a devil who sticks around with unshakable determination all your life."

Shunkichi's wife placed supper plates on the table and brought a pourer filled with heated sake. In contrast with thin Shunkichi, whose only strong features were his thick bones and his height, his short wife was plump, her face pink like a young girl's. I had heard from Naoko that she had a fine disposition. The two boys were moving excitedly around Shunkichi and me. At the frugal supper table, Shunkichi and I became inebriated enough to become somewhat glib.

"Do you believe the excuse that they used the atomic bombs to end the war?" I asked Shunkichi.

"Don't take me for a fool. Who would believe their statements and excuses about atomic power? Ask J. R. Oppenheimer if he did or did not advise the president to use atomic bombs to seize the initiative in the war. Or was it slaughter in H and N cities for the sake of experiments in the atomic era? How long will they continue to believe that their discovery and deep knowledge of atomic power have embellished a brilliant page in human history?"

After a pause, I said, "When you speak like that, even your face starts to look beautiful. Then, why are you living this way like a hermit?"

"My health's no good."

"Does your old man say the same thing?' "

"He's no good, either."

"Is he ill?"

"Ulcer, you know. He's trying to live by fleeing from this floating world. Neither my old man nor I can join the peace movement and have the same influence you do."

"That's not true. You have influence. Don't talk nonsense. If the whole society joins, of course there's influence."

Shunkichi drank sake in silence. Suddenly I said:

"Do you know about the child? Mine? If you know anything, would you tell me?"

"I don't know anything about his recent past," Shunkichi answered, gloom over his eyebrows. Being a gentle person who did not like to see another in agony, he said with deliberate humor: "When he was small, his old man over there brought him on New Year's Day every year, so we saw him, too. It was funny that he called the old man over here Dad. Once he saw our toys and wanted them. I heard there were many over there because they loved him, but he wanted ours. He especially seemed to crave my long sword. After he left, we were missing many things. The old man over here, I heard, gave them all to him!"

I was afraid of asking whether my child was alive or dead. If he had died, it was better not to learn it. If it was in vain to try to find out about the child, it seemed meaningless to stay long.

"Would Papa know?" I asked about Kurata, using the name Shunkichi and the others had used in childhood.

"I think he probably doesn't know what happened to him after the war. If he did, he would say something to us. He doesn't say anything. He is reserved about the boy's present mother, too. She's extremely afraid of the child's seeing you."

This rang a bell in the depths of my heart.

"You and even Naoko have visited me, but not my child."

I swallowed the word death, which came up to my throat.

"Kurata knows nothing, does he?"

After a while I started to rise.

"I've stayed long. I'll be off now."

"Going home? You can't."

"Why?"

"Because it's raining hard. There are no more buses, so if you're going home I'll have to take you on my bike. I'd rather not."

In deciding to stay there was a fantasy about Kurata. It was a fantasy containing lingering thoughts. Before I realized it, I had counted him among those whom I should see while in H city. Shunkichi's youthful wife, who had hung a mosquito net in the other room and was putting the children to bed, came out and started to prepare my bed in the guest room.

"This house was destroyed, too, wasn't it? I thought there was no damage up to here," I said to her.

"Yes, the walls are leaning quite a lot and crumbling everywhere. Half the ceiling blew off, and on rainy days it leaked all the time as if there were no ceiling at all. Last year at last we had just the ceiling fixed."

Shunkichi and I went out to the entrance and waited there while she hung the net. Of the bikes on the dirt floor, the red woman's bike shone. Leaning against a high desk, I said:

"You're lucky. To buy a pretty bike like that for your wife and to bike together—it's a nice life."

"I want to take good care of my wife."

Hearing this, I no longer felt that the words carried an insinuation about my past. Turning around, with no special thought I eyed the wall against which the desk stood. On it hung a *shikishi* poetry card with the words "Human passions are a void." I see, I thought, understanding Shunkichi's mood which permeated this house. Below and to the left of the calligraphy, signature-like letters said "Hansen-An," Half Saint Hut.

"What's Hansen-An?"

"The house where my old man is. Instead of a normal vertical wooden plaque with the address and name, a horizontal one saying Hansen-An hangs there."

I felt a blow.

"Is this your old man's *shikishi*?"

"Of course."

With this I held my silence. Just as had always happened in the old days, my legs suddenly began losing strength due to a profound sense of despair concerning Kurata. Long ago when I was leaving he said, "I don't want you to become happy through a man, nor will you be able to. Please find happiness in your work." Despite his socialist views, he despised me as vulgar for going to see the summer festivals, Bon dances, and other popular entertainment. He even called me vulgar for wearing *komageta,* flat clogs made of a single piece of wood, on a fair day. Now, once again I realized his illusion. That was because I thought the philosophy of Hansen-An was the illusion of his life.

I slipped inside the mosquito net. Shunkichi brought two electric lamp stands.

"What am I supposed to do with two?"

"One's a big lamp, the other's small. You're going to read the evening paper, aren't you?"

Shunkichi put the single-sheet evening newspaper inside the net.

"Thank you. But one lamp will do. D'you have any sleeping pills? I always take some."

"I'm afraid we don't have any sleeping pills. Why don't you drink some sake with egg?"

There was no way I was going to be able to sleep with something like egg wine. Shunkichi went out to the kitchen with his wife and made a clattering noise. After a while, his plump wife handed me a cup of wine with egg, laughing. In a big white cup, the hot sake and the raw egg were separate, and there was a stinging smell of sake. At midnight I still couldn't sleep. I had kept the lamp light on. The big net, dark blue from age and with stains in many places, hung limp. Here and there I saw slender white hairs stuck to the net. It was uncanny. I knew that Kurata's mother, who had lived with me long ago as my mother-in-law, died in this house a year ago. So the white hairs seemed to have belonged to Kurata's aged mother. Yet I had an illusion that they were Kurata's.

I shut off the light. It seemed to be dawn. I gathered and pulled the blanket over my eyes. When I tried to sleep, memory sprang back. Seven years ago when I escaped from H city, which had gone up in flames, and found shelter in a house in a mountain village, a man walked along the only village road saying the same thing every day.

From the railing of the second floor, I stared at that man, who was past middle age. He walked around, tenaciously spreading the rumor: "I hear that every single person who was in H city that day is going to die." He was a rich man who had moved to that village years earlier. Therefore he did not see H city on the day of the bombing. There was no basis for totally negating what he said. Even today, there was nothing that could completely bury the man's words.

As I tried to sleep, a big, vague, white hand appeared before my eyes. It was a hand that disturbed me all the time. It was the hand of the man who released the atomic bomb. That hand pressed the button, pulled the switch, causing the first atomic bomb to drop. What did he look like, the man with that hand? Above this pilot, there was an officer who commanded it. Above the officer were capitalists, statesmen, and scientists. However, I wished to see just once the face of the owner of the hand that actually dropped the atomic bomb. I wished to find an answer in his eyes to the question as to whether or not the wine he drank today in his own country was as bitter as death. Or rather, I wished to ask the soul of the man who pulled the switch with that white hand whether or not he was able to live now free of agony.

While seeing a vision of the white hand, I slept a little as it was getting light.

My stay in H city continued several more days. It was now July. I had distanced myself somewhat from thoughts of suicide and homicide. It seemed that I had started to think I should not die.

The rainy season was not over. Sometimes there was a torrent. Before and after that, it now rained and now stopped raining. In the early afternoon one day, the older of Teiko's two daughters came home in the rain, carrying a young bamboo branch. The green bamboo leaves, wet from the rain, seemed alive and breathing.

"What's the Seventh Night, *tanabata,* about?" the girl asked me.

"A cowherd star and a weaveress star meet each other on the evening of July 7 across the river in the sky called the Milky Way, after a long time. After a whole year."

"That's tomorrow night."

"I see, the bamboo branch is for that, then."

"I've got to go buy *tanabata* decorations with Mom. Everybody's going."

Teiko went out to buy decorations with the two children. My mother, who never used to nap before, was taking a nap in a corner of the six-mat tatami room. Since at night slugs only multiplied, my mother stayed up almost all night getting rid of them for Teiko, me, and her grandchildren. During the day when slugs didn't come out much, she sometimes winked. I too lay near my mother's feet. As I was starting to fall asleep, Teiko and the children came home with hurried sounds of clogs, talking cheerfully.

"I bought the eighty-yen set because they said they'd price it down to sixty yen."

Waking our mother and me, Teiko spread papers of different colors for the July 7 decorations all over the place. Bright yellow, red, or purple paper squares, gold and silver square *shikishi,* and narrow *tanzaku;* pink treasure boats, papercut eggplants, and gourds; circular and angular paperfold balloons and lanterns. . . . The children and Teiko, and also my mother, blew into the balloons and lanterns making them swell. I vaguely watched more and more swollen balloons and lanterns, then lay down again.

"Oh my, this is the Milky Way," Teiko showed the children, lightly pulling a long blue and pink paper with cuts that filled both her hands.

"We're supposed to cut these square colored papers and gold and silver papers in *tanzaku*-shaped strips, write something on them, and hang them, right? I wonder what to write."

"You're supposed to write wishes and poems," my mother responded. The older child said:

"At my friend's house, they already hung all the papers. She wrote 'Father' and 'Mother' and hung them."

"Big sister, what are we supposed to write? Do you know?"

"When I was small, grandmother used to cut pieces of cloth of different colors into *tanzaku* because there was no square colored paper in the country, write words on them, and hang them up. I wonder what words she wrote. I don't know too much about the July 7 festival."

I felt interminably sleepy then.

"All you need is to write whatever you wish and hang the papers."

Hearing my mother say this to Teiko, I said on the spur of the moment: "Write 'Against war' or something."

With that I fell asleep. When I woke, all the papers, lanterns, and treasure boats, numerous stars and balloons, were hanging from the twigs of the green bamboo which were sticking through and tied to the looped metal handle of the little chest of drawers in the three-mat room. From the top of the bamboo, a pink and light blue Milky Way streamed down left and right.

"How beautiful. Fully bedecked," I said, as if my eyes were refreshed. Unless it were the evening of July 7 according to the old agricultural calendar, one saw neither the traditional Milky Way nor the stars, so I had no particular interest in the July 7 festival celebrated in the wrong season. However, I could not deny that there was a comforting scene of a certain down-to-earth beauty before my eyes. A fragment of sad poetry was there. I noticed the writing on narrow strips of paper hanging from the green bamboo.

"Yeah?" I said, looking at Teiko. On the blue, red, gold, and silver *tanzaku,* only the words "Anti war" and "Milky Way" were written.

"Don't you have anything else to write? It doesn't have to be just 'Anti war.' "

"The children make lots of noise, and besides, I can't think of what else to write. Big sister, why don't you write, too."

I suddenly found it appealing. Taking a brush from the ink-stone box by the side of Teiko, I wrote on yellow and pale pink *tanzaku* strips. I wrote "Peace" on one, "Liberty" on another, and "Stars of love, please protect our peace" on still another, and attached them to bamboo leaves with thread. By the evening, Teiko and I had hung countless *tanzaku* from the bamboo twigs on which we wrote, half with amusement, "Sincerity," "Courage," "Stars, protect our peace," etc. At night, slugs gathered around the bamboo for the July 7 decoration and started to creep. My mother began to pick them up with throwaway wooden chopsticks and put them in a can of salt water. I mumbled to no one in particular:

"Somehow, I can't stop thinking that these slugs might be the ghosts of the soldiers who died on this parade ground."

My mother put down the can and the chopsticks in the shade of something in a corner of the room. With no context she said:

"When there's war next time, let's escape together to Miyano's, shall we? I hear there's going to be another big war."

I felt a sharp, gnawing pain in part of my chest. Again it gnawed and gnawed. It was sad that my seventy-four-year-old mother, worried that there would be another world war, was thinking about escape.

"Mother, don't think about moving to Miyano's. There's never going to be a world war."

"There really won't be?"

"No. It won't be possible even if some want to have a war."

My sister, who remained at the foot of the July 7 decoration, said, looking worried: "Even if it stops raining tomorrow and the weather's good, I don't know if we will be able to take this pretty bamboo outside. I've started to feel nervous. All our *tanzaku* have writings different from other people's."

Toward evening the following day it started to clear up. There was clear blue in the southeast sky. With her children, Teiko walked from street to street where hundreds of burned out people's houses comprised a town. After going around looking, Teiko returned breathing fast and said:

"Big sister, it's okay, it's okay."

"What's okay?"

"I saw it written. Here, there, everywhere. We must put ours out quickly. Big sister, please come out."

Wearing *komageta,* I went out with Teiko and walked, looking between the houses. Rows of shabby houses identical to Teiko's stood side by side. This was a gathering of people in bizarre situations. Only atomic bomb survivors and returnees from the front lived here. On paths between shacks that stretched as far as we could see, vegetables grew and summer flowers bloomed. Here and there under the eaves of houses in the back alleys, a green bamboo stood with July 7 decorations. Perhaps bought at the same store, pink treasure boats, gold and silver stars, circular and square balloons, and lanterns swayed in the breeze just like those of Teiko's family, together with *tanzaku* of many colors. One *tanzaku* said "Anti war." A few uncut square colored papers read "Peace, liberty, independence." "Father" was written in a child's hand on one, and "Mother" on another. Some *tanzaku* said "Milky Way" or "Tanabata Festival." At another house, writing in pen on yellow paper read:

> Inscribing a stone from a distant day
> shadow falling on the sand
> crumbling—midway between heaven and earth
> a vision of a flower.

The paper with this poem, from the epitaph of the poet from this city who had committed suicide in Tokyo, hung from the stem of the green bamboo leaves. I felt my heart stir. I stood in front of the house for a while. Although personal experiences differed, I felt deeply that I saw a certain stance of the heart flowing through the rows of these shabby sheds. Four or five houses down, *tanzaku* saying "Oh, stars of love" peeped from among many balloons, and a lovely chain of small gold, silver, blue, and red stars hung from one that said "Peace."

This was dusk of July 7, 1951.

Memory of a Night

Sata Ineko

I

It was the summer of 1951. It happened during a trip. She had a thought, which was also encountered by many others on their own journeys. The incident should eventually clarify itself. She recalls, simply as an experience, what she faced on that trip. It concerned the Party—meaning, of course, the Communist Party. She was a Party member. At least in her mind she was still one then. And she was an author.

That summer, at the invitation of a labor union, she was on the way from Tokyo to the summer cultural series at a T Prefecture factory as one of the lecturers. She rode the train, as directed by the union, to N terminal on the coast of the Japan Sea and transferred. A connecting train was waiting at the same platform. She assumed without thinking that it was her train. The train stopped at each station, unloading people. Few got on and the seats around her gradually emptied. The train had already run more than two hours. Finally she began to look suspiciously at the landscape along the railroad. The train was supposed to run along the Japan Sea. She had wished to see the gray, heavily swollen ocean again that day outside the window. However, the train never came to the coast. Something was wrong. Wearing glasses, in an indigo cotton blouse, and middle-aged, she alone looked out of place on this train which dropped people at every little station. Feeling restlessly that

---

Yoru no kioku (1955). Translated by Kyoko Iriye Selden with the permission of the author.

something was wrong, she felt isolated, a stranger. Somewhat flustered, she took out the letter that had come from the union to check her route from the starting point. It was a businesslike letter written over carbon paper. As expected in such a letter, the train schedule and the transfer point were written in clearly. It seemed that she had taken the trains as instructed. However, while poring over the letter she noticed an error. She had taken the train at N station following the signboard on the platform. The name of the station on it was, however, different by one letter from her actual destination. She had hastily identified the station and traveled in the opposite direction. No wonder the train never reached the coast.

She quickly got off at the next stop. It was a quiet, small station. As she explained what had happened at the ticket gate, the young attendant, relaxed after allowing several passengers to exit, looked up as if at a loss, as much as to say, "Well?" There was no other train that day going to her destination, he told her. The only thing to do was to return to N station, he explained, and take the first train the next morning. Fortunately, she had no plans for that evening at the destination. In any case, there was nothing she could do but go back to N. She had to wait about fifty minutes until the N-bound train came. Though it was not yet dusk, there was no place to go from that little station. A low hill right in front of the station obstructed the view. Beyond the tracks was a farm bathed in the pale yellow evening sunlight, in a hush after the wind had stopped. Since she was by nature unperturbed by having to wait a little while traveling, that was all right; yet, thinking that she was there by mistake, and in a place which had no connection to her, she felt as if she had come strangely far. However, she would be back in N city after a three-hour ride. She knew a family near the station. Maybe she should not, however, stop by there. All she had to do, she thought, was spend one night at an inn in front of the station. After being seated on the N-bound train for a while, she still thought about the inn.

When she arrived at N, it was already night, which made her all the more aware that she had unexpectedly wasted time. Moreover, it was drizzling. In spite of the fact that N was a fairly large city, the front of the station was quiet and traditional, with no tourist atmosphere. The city lay just beyond the narrow square, with rows of shops. Standing in front of the station, she saw an inn right across the square. It was a decent place. There was a humbler inn beyond it. Any place would do. But if she walked a little along the street running vertically from the station front, there was the house of her acquaintances. It was so close. On both occasions when she had come to this city before, she had stayed there. This time, however, her visit might inconvenience them.

Would the couple perhaps refuse her? It was easier for her to imagine that she would bother them than that they would refuse her. This was due to her feeling toward them. The husband, purged from his job, was a worker type who gave a sober impression. The wife, too, larger than her husband, had a character drawn, so to speak, with a dark pencil, far from anything like the stereotyped gently smiling female countenance. In the deliberate yet brisk way she dealt with matters she betrayed the wisdom of one who battled for a livelihood. They had five children, the oldest girl with bobbed hair. Of course the husband was a member, and the wife, too, no doubt was another. Thus the atmosphere of the Party pervaded this house.

Was the green grocery continuing to do all right, she wondered, visualizing the household which, on her second visit, had just started to sell dried food, preserved food, and vegetables for its livelihood. When her mom and dad went out to sell vegetables from a cart, the eldest daughter was her mom's right hand, taking care of the shop and her brothers and sisters. She wondered how the girl was doing. Reticent like her father, her eyes tended to be cast down under her long bangs, but she was a good-natured hard worker. The girl had silently imbibed the Party atmosphere of the house, and wrote poems in that vein.

She liked that family. Rarely was an entire home so perfectly open toward the Party. Elsewhere there tended to lurk an uneasy facade or hidden incoherences. With a jobless husband and five children this family was obviously poor, but everybody worked hard to fight through the poverty. If there were minor flaws, even those were undoubtedly of the kind natural to workers.

Therefore she did not doubt that the family was still involved in the movement while running a green grocery. That was why she hesitated, thinking that her visit tonight might be an imposition. She had stayed there during both of her two previous visits to this area. How would it be for her to stay at an inn without even saying hello to them? A dark thought swelling in her mind, she questioned herself: How would that be? After a while, though feeling anxious, with her suitcase she started to walk in the rain toward the house, prompted by trust and love for the family.

II

Though the glass doors were half closed, the lights were still on in the shop. One step in on the dirt floor, she stood hesitantly.

"My," the wife said slowly in a deep voice, gazing at her. As for her, she instantly groped for something in the wife's expression.

"How unexpected. What happened?"

"I'm sorry it's so sudden."

"Come in anyway. Our place is always a mess—"

The dirt floor connected the shop to the kitchen inside, and without even a step, there was the dining room, and the cash register was also there. When she got in, the children came running from inside. The eldest daughter, who was doing something on the dirt floor of the kitchen, faced her and bowed, smiling faintly. She sat close to the entrance, and after re-exchanging greetings told them about the day's incident. Then she said,

"Maybe it's wrong to come to your place. Forgive me. I just didn't feel like going to an inn and ignoring you. But I'm afraid I may be troubling you. If so, I'll leave right away. You know, don't you, about me lately?"

"Yeah. We didn't understand it well and were wondering what it was all about."

Pouring tea with her eyes down, she added, "Let's see, Dad happens to have gone out to stock up today and won't come back till tomorrow morning. But, since you've come all this way—"

Hesitation surfacing within her breast, the wife's expression had stiffened before she knew it, while maintaining the usual bold lineation. It showed her anxiety about the Party rather than toward the guest, who recognized it and found it natural.

"Well, it would be wrong to ask to stay. I knew it, but—"

"Oh, no, of course, you know— We don't feel any differently toward you, but— A young district committee member is staying with us, you know. I don't know what he'll say. Whatever he does say, do you want to stay until then anyway? It's good of you to come, and I don't feel like asking you to leave—"

While they were talking this way, a neighborhood woman came to buy *miso* paste for the next day's breakfast, and a young bachelor came to buy preserved food. While handling these sales, the oldest daughter also listened to the conversation between the mother and the guest. Not only that, it was clear from the way she moved this way and that on the dirt floor that she was probably concerned with how the matter would turn out but was disposed in a friendly way toward the guest.

"Will it be all right? I don't want them to say anything about you. That's what I'm worried about. Shall I leave? Somehow I couldn't come here and not drop in—"

While speaking, she felt as if they were carrying on a conversation in a stale melodrama. Neither the wife nor herself, while implying a great deal, directly referred to the affair that had caused this situation. That the wife's expression betrayed an inner struggle but not hatred toward her might have corresponded to the stereotyped quality of the traditional melodrama. At the same time, of course, there was something sentimental about her feeling toward this family.

"It was so good of you to come. I'm happy you thought of coming, but then—it's somehow too hard for the likes of me to understand. But, well, anyway, why don't you stay for now. If it's really something, there are inns nearby—"

"Is it all right? Somehow I don't feel like going to an inn—"

"There's no reason to go to an inn when my house is here. Have you eaten?"

"I had a box lunch, so—"

"Then please have some tea. Yasuko, peel an apple from the store for her."

In a way that expressed relief, the oldest daughter went to the store right away to get the fruit without answering.

"Well then, please relax. I've got to wash the carrots."

"I'm glad that the shop seems to be doing gradually better."

"Well, you know, with a family this size, livelihood's hard. Dad can't be very active in the Party."

"It must be really hard."

"Me too. There's talk about organizing women, but the situation's getting more and more difficult, you know. It's hard to move. Though it's not that I'm doing that much—" she said, standing up to go to the kitchen. "Where did you say you're going this time?"

"I'm invited to a spinning factory. I think it will be a learning experience for me, too, to meet young spinners."

"I see. You're busy too. Well, please give a good talk to the young spinners."

"I'm the one who wants to hear them talk."

"Even so, they'll all listen to lecturers from Tokyo. Local communities need that kind of thing. Your last visit was for a cultural lecture series."

As the wife had made up her mind about the guest's staying, and the guest herself felt set on it, the two started to talk animatedly. That eased her restlessness, and she shared with the children the apple the oldest daughter peeled for her. The youngest, a boy, had been a baby on her last visit. He was already three or four. That meant that she was here after three years. The boy, like neither his father nor his mother, smiled all the time, and though he didn't talk much, he ran around the rare guest moving his round eyes as if he just couldn't sit still.

"It was bad of me not to bring a gift. And I have only one of these." She held out the last boiled egg from her bag. He took it, looking surprised, and went to show it to the second daughter, an elementary school student with an open textbook in front of her.

"Oh, Mom!" The big sister in elementary school took it from her brother, reporting it to their mother, and added, "We'll all share it, all right? I'll slice it for you."

The big brothers, also studying, glanced at it. The little brother eagerly gazed at his sister peeling the egg with his round, impish, lovable eyes, without saying anything. With his hands on the edge of the desk, he stamped his feet repeatedly.

The second oldest daughter, having carefully peeled the egg, said, "Wait," and brought a cooking knife. The egg rolled under the big knife. But the girl neatly cut it into four pieces. Since she sliced it crosswise, the two corner pieces had little yolk. That didn't seem to matter. Anyway the single egg could now be shared by the four children.

"Look, I cut it."

The older brothers' hands, too, stretched out. They each picked up a piece and tossed it into their mouths. The youngest child remained looking at the piece of boiled egg in his fingers as if it were a treasure, and smiled at the guest. The second oldest daughter, after putting her portion into her mouth, returned to prepare for school, satisfied that she had completed the task of distribution. The single boiled egg was shared by the four children without a complaint, disappearing into their mouths in a second. As if contented, the older children studied, while the youngest was somehow still interested in the guest. He sneaked up from behind and put his arms around her shoulders as though to hug her tight. At first he appeared hesitant, not sure if he should do such a thing: then he peeked at her face and smiled. The oldest daughter, who with her mother was washing the carrots to be arrayed in the store, went out front each time a customer came in.

"Yes, that's right. Thank you very much."

She was blunt, but she didn't forget a polite thank you. There were eggs in the store too. "To organize women," the mother had said a while ago, but now she was washing dirt off the carrots, her sleeves rolled up to expose round arms, while instructing her daughter on something about the store. This housewife with the robust hips also talked in front of people at various Party meetings.

Wouldn't the wife be criticized for asking her to stay? The guest felt as if something was going to happen that very evening and yet she was there because she had a feeling of love and trust toward the family. But what did that mean? The reason that she dared, so to speak, to sit there was owing to this doubt. Rather than doubt, it might have been confidence. Confidence in her feeling of love and trust toward this family—it might have been something like that. Without that, she couldn't have sat there. This house, she knew, was not only the residence of the family members, but also a lodging for Party activists. Moreover, it tended to be a site for Party activities.

Rising slowly within her now, like a wave, was the tremor that had risen at that time in the past from the bottom of her breast, draining her entire body. Whenever she recalled that moment, that sensation in her entire body came back. That time, it was deep night. It was in the suburbs of Tokyo, where many fields and woods were still left. She crouched on the road alone by the side of the woods. Beyond the road were several city-run apartments, from which she had just come. Though she had come outside, she was unable to walk, and remained crouching there. Her inability to walk was not just owing to fatigue. It was already past twelve, but people were still in that house. After an all-night debate, she had just been driven out of the cell on their decision to expel her. So she crouched by the roadside, unable to simply return home. The road near the woods had few passersby even during the day; so no one could be expected to pass. The few city-run apartments in front had grown dark, and it was almost odder that a woman was crouching there than it would have been if the place had been totally deserted. A tepid late April wind blew from time to time carrying the smell of soil.

Was I ever meant to experience this in my life? she asked herself, lowering her shoulders in a complex sense of sorrow. It wasn't the kind of sorrow that made her cry; rather, her entire body became void, and a shiver rose from the base of her breast. In addition to the dozen or so cell members, many of whom she had had frequent contact with, there was also a young farmer she didn't know and a young man from on high who was present that day especially in connection with her punishment. She met these two district committee members for the first time.

Her expulsion was a result of the sectarianism of those days. However, in her case, which involved nothing but a difference of opinion, there was an atmosphere didn't easily allow a decision for that reason alone. Among those she had contact with were one or two close friends. Not because of that, but because the case for expulsion was weak, the unknown village youth joined in insisting on deferring the decision. The district committee member cornered those with soft-hearted views. So long as a directive from above called on them to "examine the thought of the concerned party and punish her," the young district committee member was determined, as his ominous expression revealed, not only to "examine" her thought but to "punish" her. A despicable fag end of a sect—this, it seemed, was she. Moreover, she was a habitually insolent novelist, he seemed to think subjectively, which made him even more ferocious. He was lean and pale with drawn cheeks. He thrust out his arm to point at her, opening his mouth wide to spew forth harsh words. Irregular teeth were exposed in a sickly blackish mouth. He rebuked her for irreverently grinning. She didn't at all mean to deride the debate concerning herself. On the contrary, she was seriously angry, even sad. And yet the district committee member seemed to whet his irritation and authority by arrogantly shouting that she was grinning. The air in the room was oppressive. Only the young cell leader, as though to prove himself to the higher-ups, joined the district committee member in enumerating her accustomed faults. Her attendance at cell meetings was poor, she took the liberty of holding regular gatherings for literary associates without reporting to the cell, and so on. Such arguments sounded shallow and forced and only served to make the atmosphere of the meeting unbearable. The reasons for expulsion were not strong enough to convince all the cell members. At last an elderly historian resorted to something blatantly transparent:

"Can we establish the fact, for example, that she has pocketed some money?"

This was not only an insult but was so unexpected that no one even answered. No one even whispered an aside in this debate, out of which the higher-ups seemed determined to expel at least one person. Twice, three times, they were forced by the district committee member to state their views one by one, and as this was repeated twice, three times, the number of those arguing for deferring the decision perforce decreased. For they themselves were vehemently criticized for their soft compromising spirit. Those who opposed expulsion to the end were just the three who were close to her, and the expulsion was resolved by a majority vote.

"I cannot accept it, never."

What she was allowed to say during the debate was turned by the district committee member into material for exposing her before the cell members, since it did nothing but prove her insolent sectarian thinking. She kept her eyes wide open, resisting a feeling which might be called despair at the stupidity of it all. At the end of the debate, she just had to say that she could never accept the decision.

"Ah," the district committee member said contemptuously, "it's possible for you to submit a protest to the general meeting."

And she was ordered out before the dark expression on the faces of several cell members. Crouching close by the house was her act of resistance. She sat down on a small rock, an elbow on her lap, resting her cheek on one hand. Something was wrong about this situation, something was wrong somewhere, she thought, ruminating with an odd calm on the condemnation that had befallen her. Partly she laughed at herself, realizing that, however unexpected, such a thing had really happened to her. But something was wrong somewhere. Something that made her say she could "never accept it" rescued her from her self-scorn.

She peered at the dogs appearing under the lamp at the street corner. They flocked together, as if liberated, on the midnight street. Nearly ten of them came toward her, sporting with each other. They had not noticed a human being crouched on the street. Since she thought she must look suspicious crouching there, she readied herself against the dogs, peering at them. They were now chasing, now being chased, in a good mood and not noticing her until one of them came close to her knees and flinched. Taken aback upon glimpsing her, it retreated a step, as much as to say, "what a surprise." "Oh, just a human being"—it seemed relieved and ran away joyfully as another dog sprang on it.

One dog ignoring her, the others also paid no attention, going back and forth before her in a group. The footsteps of the several dogs of various sizes made mincing four-footed whispers. As though reverting to their wild instincts, they sported on the midnight street for a long time. There was no other sound in front of the woods. Only the footsteps of the four-footers continued. She suddenly felt sorry for herself, crouched near them without budging, and after a while she rose to go.

III

The children were put to sleep; a curtain was drawn across the glass doors of the store.

"Why don't you sleep now? We stir early in the morning," the wife said, looking up at the clock. "I'll talk to him, all right?" she said, lowering her voice at the end of the sentence.

The district committee member staying upstairs was expected back late. At this suggestion, she went to the bed prepared for her upstairs, but for a while she struggled to suppress the memory of that night which started to move in the bottom of her heart, stirred by the atmosphere of this room. There was only one room upstairs. There was a small desk. Nothing else. Her bedding was spread close to the wall on one side. The district committee member shortly to return was to sleep near the wall across from her. From previous experience, she knew her way around. Staying over itself was already a problem—now she was to sleep in the same room with the committee member. Under the dim electric light, she imagined his return. District committee member—he must be a sharp fighter. When that district committee member with drawn cheeks had opened his mouth as if howling, the inside of his mouth had looked blackish and sick. A thought occurred to her: had his lungs been infected? But why were there such human beings? A type: lean, pale, sharp-tongued, honed like a razor, with shallow arrogance as if finding pleasure in thrusting an arm straight out to point at a person and smash him. It was a type.

She regretted that her memory of him was only as a type. Although she remembered him clearly as a certain type, she was not confident that she would be able to recognize him if she met him somewhere again. Her hatred from that night was, in other words, directed only toward a type.

She had begun to doze off before she knew it, when she opened her eyes at the voices downstairs. They must be talking about her. The wife's voice was heavy and somewhat muffled, and the man, too, talked as if forcing himself to hold down his voice. I'll be called down, she thought, and just then heard footsteps on the staircase.

"Say," the wife said hesitantly, "are you awake? Listen, I'm sorry, but he says you're to come down."

"I'm coming right now," she answered, implying that she knew what it was about, and getting up. As she went down the steps, a young man seated near the foot of the staircase with his face down looked up.

"Hi," she greeted him face to face. He started without introduction, as much as to say that he had called her in order to tell what was to be told: "For you, you see, to stay here is, you see, partly for the reason that this is not just a private residence, you see, though I don't know how you think of it, from our standpoint, troublesome."

The young district committee member, owing to the character of the problem, sounded antagonistic. However, he was not arrogant. Grew up in poverty, became a worker—that was the impression he gave, a character spotted everywhere and yet not particularly conspicuous. He looked like a man who would pursue his job sincerely without minding what went on around him. He examined her with that sincerity. The wife had left to join the children.

"When I'm asked what I think of it, I don't know how to answer, but I did think about whether it would cause trouble, yet— It was hard to come here but not drop in."

Her answer wasn't at all logical. The young man looked as if he had missed the target.

"However, you see," he pressed, "you're walking along a different road from ours, aren't you? Suppose a person like that, you see, stays in this house, what would that mean?"

"What different road? I'm not walking along a different road."

"Well, but, you people are already engaging in activities hostile toward the Party, aren't you?"

"Against the Party—?" she lifted her face as though to question back. "I don't think that we engage in hostile activities against the Party. This involves, I think, various problems. There are many factors."

"I don't, you see, recognize the need for developing a theoretical struggle with you, but in any case, what is the purpose of your trip this time?"

It was nothing but an inquisition. For him, she might have been an enemy spy. On hearing the purpose of her trip, he still imagined it in direct connection with the organization.

"Then, you will go to the factory and on your return report to the higher-ups?"

"There's nothing in particular to report to higher-ups. Due to the nature of my work, occasions arise when I'm invited somewhere, and if so, I talk from my own standpoint."

"Oh, is that right."

He was thinking to himself, looking as if somehow he didn't understand it well. And now that something he could not understand surfaced, he seemed to realize that the situation was different from what he had comprehended. He said, "Well, in any case, you see, now, at this time of night, probably I can't ask you to leave, so it can't be helped. But tomorrow morning, I expect you to leave early?"

"Thank you very much. I'm sorry to have troubled you."

The wife, who had lain by the side of her children, might have anticipated this conclusion beforehand. Nevertheless, she must have strained her ear to follow the process. Climbing the steps, she felt simple-heartedly warm toward the young district committee member who had brought this around to its natural conclusion. Slipping under the covers, facing the wall, she felt that this was logical enough for her in its own way. And yet, even so, what did it mean that they, namely the housewife, the young committee member, and herself, all had to experience such delicate pain? Something was wrong somewhere—she could only think so. The footseps approached, mounting the stairs. The young district committee member, since this room was his sleeping place, had no choice but to prepare his bed, ignoring her. She heard behind her back the sounds of his quick preparation. And soon she too fell asleep.

In the morning she rose quietly. Downstairs, the wife and the oldest daughter were already up.

"Oh no, I didn't have to eat," she said, seeing that breakfast was ready.

"But please, at least do that before you leave," the wife said.

There were eggs on the table. The daughter occasionally smiled modestly, but seemed to hold back from talking to the guest. She, too, knew of the problem within the Party.

Saying that it was so close, the wife accompanied the guest, despite her protest, to the station.

"Young people, you know, are single-minded, so—," the wife said, by way of excusing what had happened the night before.

"Well, he probably had to say what he did, so I don't mind at all. Rather, I feel sorry about bothering him," she answered. In front of the station early in the morning, commuters to the factories were hurrying along, and the station shops were beginning to open.

"Well of course he has responsibilities, so perhaps he can't help it."

"I'm sorry this happened when your husband wasn't home. Please give him my regards."

"You may meet him at the station. He's due on the same train that you are taking."

As she apparently intended, she went onto the platform with her. After a while the train arrived, and as she got on, the wife spotted her husband and brought him near the window where she was looking out. It seemed that she had hurriedly informed him of the events of the night before, for the husband responded to her greeting by saying, "Yeah, yeah," with a vague expression, keeping his distance as though at a loss. He had gone to buy vegetables in the same work clothes he would wear when pulling a cart.

Since there was a little time before the train started, the couple stood before her window, but the husband remained silent as if he didn't know what to say. It was because of her delicate situation, as was evident in his distant expression. The wife, too, seemed to understand it. The wife said at the window, as if to both the guest and her husband, "Since we like you as a human being, so— Right, Dad? So, if you come this way, please stop by again."

This must have been said from something like relief at the fact that the guest was finally leaving. Letting the words "we like you as a human being" linger in her ear and beset by complex feelings, she said, "Thank you very much, but forgive me if anything annoying happens later."

"I trust that it'll be all right," the wife said.

When the bell rang for the train to start, the husband, too, for the first time recovered his usual expression, saying in a clear voice, "Take care now." It sounded as if, since this was the moment the train was leaving, he had put aside the Party problem just for that instant. By doing so, he seemed to feel relieved himself.

After dropping off people going to work in N city, the train was filled with the voices of local people mingling with each other. Near them she warmly recalled the delicate psychology of those from whom she had just parted; however, the warm feeling, at this point, gradually changed to pain. She soon grew tense and started to think somewhat farcical the course of events since last night which placed everyone including herself in an awkward situation.

Today, the train soon ran along the coast of the Japan Sea. The ocean spread in the exact same color as the clouded sky. Somewhat farcical—thinking so, and yet owing to the forced character of the thought, she gazed at the color of the sea with an expression of sorrow.

Love in Two Lives: The Remnant

Enchi Fumiko

Kneeling down in the open corridor, I called toward the inside of the *shōji* screen, uneven in color because of the patched repairs which had been made here and there,

"Professor, may I come in?"

A cloudy, ambiguous sound which might have been yes or no came from inside, and I heard the dull sound of the nightcover sliding to the side as he changed position while lying down. Since I expected this response, I opened the *shōji* quietly and entered the room in my overcoat.

Just as I had imagined, Professor Nunokawa was searching for the revised edition of the textbook, thin and large format, which was lying at the side of the bedding, slightly raising his head with dishevelled white hair from the soiled pillow. Each time I came here to do copying for him, the nappy, soiled sheets and white calico collars of the nightcovers bothered me, but Mineko, the maid who took care of the professor, showed no intention of changing them. The dirty bed of a sick person even if he is a young man makes one feel wretched enough, but in the case of an old man it is all the worse. My feeling of sorrow or pity had long since changed into disgust at the wretchedness displayed. Although the moldy smell of the sick man's room heightened my feeling of disgust, I asked after his condition gently as I opened the notebook for copying on the old, completely lusterless rosewood desk drawn up close to the bed, probably by

---

Nise no en—shūi (1958). Translated by Noriko Mizuta Lippit with the permission of the author.

Mineko before she went out shopping. While a bed warmer was placed in the professor's bed, the few hard charcoals in the brazier were always half-extinguished in this room, and on a day like this, when snow mixed with rain was falling, the room was stone-cold. As Professor Nunokawa had been kind enough to tell me on my first visit not to take off my coat, I continued to sit down without taking it off, accepting his offer.

"Today, we are supposed to do 'Love in Two Lives,' aren't we?"

He did not seem to want to talk about his condition, and opening the thin book on his chest, holding a red pencil in his right hand, and moving only his eyes under the thick frames of his long-distance glasses, he looked toward me. Opening on the desk the same revised edition of *Tales of the Spring Rain,* I said,

"Page 59, 'Love in Two Lives,' from the beginning, sir."

I had taken up the task of writing down Professor Nunokawa's oral translations of Ueda Akinari's *Tales of the Rain and the Moon* and *Tales of the Spring Rain,* for he was too ill to write them down himself. The publisher for whom I work is planning to publish in one volume his translations of the works into modern Japanese. Despite his illness, he was enthusiastic about undertaking this work, probably because of the necessity to earn a living. I had already finished writing down the nine grotesque stories of *Tales of the Rain and the Moon* and the first four episodes of *Tales of the Spring Rain,* copying down the words which came from loosened lips covering uneven teeth, flowing continuously like silk thread emerging from a silkworm.

Tales of the Spring Rain was written by Ueda Akinari in his later years. In the preface he writes,

> Spring rain continues to fall today. Quiet and moving. I took out the usual ink slab but after thinking awhile felt there was nothing to say. I am trying to write *monogatari* in the classical style. Yet since my life is as base as that of a mountain peasant, what can I narrate? I have taken what has been written in the past and present as true, and now it is my turn to deceive others. Be that as it may, while I continue to narrate, thinking to myself that there have been some who have forced their own fictions upon people as actual history, the spring rain falls continuously.

Akinari, who rejected as the style of a novice the world of *monogatari* he had once woven exquisitely in *Tales of the Rain and the Moon,* was still consumed by an engorged untamed passion in his later days. In *Tales of the Spring Rain,* he let that passion flow as freely as possible into historical personages and oral and folktales, creating stories which went beyond the moral framework of the feudal period. It seems natural therefore that this work, unlike *Tales of the Rain and the Moon,* has never been published or even circulated widely as a private copy. In his later years, Akinari, who had no children, lost his wife and the eyesight in his left eye. He had to endure an impoverished life for a long time in a world of dusk.

The situation of Professor Nunokawa, a famed scholar of Edo literature, was similar in certain respects. His son had been killed in the war, his wife had left him, and his married daughter, reacting against his uncompromising character and Mineko's presence, rarely visited him. Since he lacked a pension or retirement plan, a few of his students helped him by arranging to have him edit textbooks or make oral translations like this. Although I did not notice while translating *Tales of the Rain and the Moon,* I often felt after we moved to *Spring Rain* that there existed a close bond between Akinari's later years and those of Professor Nunokawa. I often felt that the words I heard were Professor Nunokawa's own, coming with a natural resonance from the depth of his inner world.

Professor Nunokawa began narrating slowly in a low, suppressed voice, as if starting to recite the sutras. He read lying down, the book resting on his chest.

It was autumn in the country of Yamashiro. The leaves of the *takatsuki* trees had all fallen down, the wind blew through the villages, and everywhere it was desolate. In the village of Kosobe lived a rich farmer who had been settled there for a long time. Owning an expanse of mountain and fields, he lived comfortably without worrying each year whether the crop would be good or bad. He learned naturally, therefore, to love reading and spent every day reading until late at night, without seeking the company of any friend in this mountain village. His mother worried about this habit of his and used to tell him, "Now, you must go to bed. The midnight bell has long since rung. Your father used to say that if one reads past midnight, one will be exhausted and surely end up becoming ill. One will go too deeply into the things one likes without realizing it only to regret it later."

He took his mother's scolding gratefully as coming from her parental love and tried to go to bed soon after ten o'clock. One night, when a slight rain was falling continuously and not a single sound was heard in the pervasive quietude, without realizing it he spent more time than usual reading. Thinking he had forgotten his mother's words that night and that it might already be after two o'clock in the morning, he opened the window. The evening rain had stopped, there was no wind, and the late moon was hanging in mid sky. "Ah, the sight of the quiet, deep night. Let me express this sentiment in waka poems—" Thinking of one or two waka poems while preparing the sumi ink and holding the brush, he noticed the sound of a gong blending with what he had first thought were only the cries of insects. Then he realized that this was not the only night that he had heard that sound of a gong. He thought it strange that he had come to notice it for the first time that night even though he must have heard it while reading books every night. Walking in the garden trying to find the place where the sound was coming from, he spotted a place in the corner of the garden where the grass was usually left uncut. The sound seemed to be coming from under the stone there. After satisfying himself that this was indeed its source, the master returned to his bedroom.

The next day he gathered his servants and ordered them to dig there under the stone. After the servants had dug down about three feet, one of the hoes hit a big stone. Under the stone, they found what appeared to be a coffin with a stone lid. When all of them together lifted the heavy lid, they found inside something indistinguishable, occasionally hitting a gong held in its hand. The master timidly came near to take a closer look at it. It looked like a man and it did not; it was bony and as dry as a dried salmon, but its hair had grown down to its knees. When a strong servant was let into the coffin to take it out carefully, he said in a loud voice,

"It is very light, very light. It feels like nothing. It is not even like an old man."

Even while people were making a fuss over taking him out, he did not stop hitting the gong. Seeing this, the master pressed his palms together with respect and said to everyone,

"This must mean he has entered *jō*, death through Zen meditation, one of the ways to attain nirvana in Buddhism. One does so by entering the coffin alive, and then dying while sitting in Zen meditation. He must surely have done this. Since I have not heard of one of our ancestors doing so during the hundred-odd years when our family lived here, this must have happened before our ancestors settled here. His soul must have entered heaven already, leaving only his body, unrotten. The hand still moving, hitting the gong as it used to do, shows his obsession. Anyhow, since we dug him out, now let us try to revive him."

The master helped the servants carry the figure, dried up and hard like a wooden statue, inside the house.

"Be careful. Don't break it against the edge of the pillar." They carried it as if carrying something fragile, bringing it finally to a room inside the house. The master covered it with a nightcover and brought a cup of warm water to wet its dried lips. Then something black, looking like a tongue, wiggled out between them, licking the lips; it eagerly sucked the water soaked in a piece of cotton.

Seeing this, women and children screamed for the first time.

"Horrible, horrible, it's a ghost, a monster." They ran away and never came near it. Seeing this, however, the master was encouraged and treated this dried-up thing all the more carefully. His mother too, therefore, joined him and never neglected to recite a sutra when she gave him water.

Fifty or so days passed, and his salted, salmonlike face and limbs gradually became moistened, and it seemed as if his body had regained its warmth.

"He will certainly regain consciousness."

Encouraged, the master continued to take care of him, and he finally opened his eyes. He moved his eyes toward the light but did not seem to see clearly. He tasted the porridge, moving his tongue when it was poured into his mouth, just the way ordinary people would. The wrinkles which resembled the skin of an old tree became shallower, while his limbs, taking on flesh, gradually began to move freely. It seemed that he could hear, for his naked body trembled at the sound of the north wind.

When the master handed him some old clothes, he took them, very pleased, and he now ate very well. At first the master did not give him meat or fish, thinking that the man was the reincarnation of a great priest. But looking at others eat such foods, he twisted his nose, showing that he wanted them very much. Seeing the man eat meat to the bone and finish off devouring the head of the fish when they were added to his tray, the master felt his enthusiasm cool.

When the master asked him politely,

"Since you are a specially fated person who entered into *jō,* death through Zen meditation, and was revived, please tell me how you were able to maintain yourself alive so long underground."

The man just shook his head saying,

"I don't know anything about it."

Then he resumed gazing vacantly at the master.

"But can't you at least recall the time when you went underground? What were you called before?"

He looked as if he really did not know anything, and in the way he retreated, hesitatingly licking his fingers, he was just like the ordinary, dumb farmers around there.

Seeing this, the master was very disappointed, for he thought he had revived a high priest after his care of several months. The master began to let him work like other servants, having him sweep the garden and spray water around. The man worked willingly without showing any dislike of such jobs.

"How meaningless the teaching of Buddha is. The religious faith which let him undertake Zen meditation and stay underground hitting the gong for more than a hundred years has gone away completely. All profundity is gone, and only the skeleton of the body has been revived. What does this mean?"

Not only the master but other intelligent people in the village as well talked in this way, frowning.

"Let us stop here awhile." Professor Nunokawa, having turned on his side without my noticing it, put down the book as if tired.

"You must be tired. Shall I make tea?"

"No."

Pursing his lips as if tasting something bitter, he said,

"Has Mineko returned? I'd like to use the lavatory. Will you call her?"

Hurriedly I stood up, and opening the paper door, I called loudly to Mineko who seemed to be in the kitchen, having returned from shopping.

"Mineko-san, Mineko-san, lavatory please."

Suffering from a malfunction of the bladder, Professor Nunokawa could not urinate well and had to use a special attachment. There was once a time, however, when suddenly feeling the urge while dictating, he had made a mistake. Remembering this I was a bit flustered.

As plump and fair-skinned Mineko, who was said to have been nicknamed Princess Usume [Princess Narrow Eyes] by the professor because of her narrow eyes, came rushing into the room, I went out to the adjacent living room.

A red, half-knitted sweater had been left on the *kotatsu* cover of printed cotton, with several knitting needles protruding from it. Mineko had evidently been knitting there until a moment before. Since it was bitterly cold, I put my hands under the *kotatsu* cover and tried to catch what was going on in the next room. Mineko was probably placing the urinal under his body.

"Ei, a little bit more, raise your hip a bit more, yes, good."

While panting with her exertion and commands, she asked blatantly,

"Professor, isn't it time to let Mrs. Noritake go back? Isn't it? Isn't it time—?"

"No, not yet. We are having a break. Anyway, let's finish this here."

"Please take your time. I will be going over here what we have done," I called out.

The professor did not answer me, but as it seemed to hurt him when the narrow tube was applied, his scolding, moaning voice could be heard a few times saying "Ouch" and "Be gentle." When his cries subsided, the faint sound of urine dripping from the tube to the urinal reached me chillingly, as if to indicate the shortness of his remaining life.

It had been over ten years earlier. Soon after I graduated from a women's college, the professor had paid special attention to me, lending me books and letting me assist in his research. With amazing boldness, he tried to touch me at that time. He had grasped my hands and even demanded more intimate contact indiscreetly. Since I was engaged at that time to my husband, who later died in the war, with the wedding day drawing near, I disdained his advances as the shameless conduct of a middle-aged man. Now that I think of it, in those days the professor must have been in the prime of manhood, full of energy to the point that he sometimes behaved scandalously—as a lascivious man whose conduct was unworthy of an educator.

In those days he had nicknamed me Princess Tamakatsura, an intellectual beauty who was one of the few princesses to refuse the love of the shining Prince Genji. After less than a year and a half of marriage, I lost my husband. A technical officer in the navy, he was killed in an air raid at the airport, and I lived the ten years of the postwar period as a wretched war widow, barely surviving while supporting my young son. A working woman living alone in the degenerate postwar society on many occasions was approached by men much more openly than Professor Nunokawa. But as the saying goes, widows in their twenties can remain widows, and the year and a half's contact with my husband kept me alive like a moistened flower, and luckily or unluckily, I have spent these years without being united with a second man. Passing thirty years of age and working at the office every day, I may appear to others dried-up in body and heart, becoming exactly like the man in the story, dried-up like a dry salmon. But in me such miracles as loving my husband in a dream or finding his face vividly in my son's face visit me continuously. Because of this, I came to regard the uncontrollable sexual advances of men in general with a feeling of shared sadness with them. Seeing Professor Nunokawa, who had persistently pursued me with such vulgar energy, make such an effort to live just trying to urinate, I was moved to tears.

When I was called into the room again, Mineko had taken the urinal and gone behind the paper door. The professor, it may have been my imagination, looked a bit more spirited, holding his face in his hand, his elbow on his pillow.

"What do you think of this story? Isn't it an interesting one?" he said excitedly.

"Indeed—I did not know there was such a story in *Spring Rain.* Is it based on some old tale?"

"That's right."

The learned professor told me of a story called "An Obsession with Meditation," in a book called *Tea Stories of an Old Woman;* it is the story on which Akinari's tale supposedly is based. The story is about the priest Keitatsu of the Temple Seikan in Myōtsūzan in Yamato Province, and took place in the first year of Shōō. When just about to begin Zen meditation, the priest became infatuated with a beauty and failed to achieve nirvana. He continued to strike the gong without his soul leaving him for the next fifty-five years, until the third year of Hōnen.

"Since *Tea Stories of an Old Woman* has a preface written in the beginning of Kampō, it must have been written when Akinari was a small child. But considering the situation in those days, it seems likely that he did not read it until some decades later. Akinari of *The Rain and the Moon* period would certainly have written in much more obsessive detail the story of the attachment to a beauty remaining behind—"

Thinking that his words also came from the depths of his heart as his own feeling, I lowered my eyes as I listened.

"There is still another anecdote about this tale. In the Meiji period, a book appeared entitled *A History of the Novel: Biographical Studies of Authors;* it was coauthored by Tsubouchi Shōyō and Mizutani Futō. It quotes Aeba Kōson's account of a man who claimed to have seen the manuscript of a tale by Akinari, entitled *Tale of a Rainy Night.* This tale is similar to our 'Love in Two Lives,' but its ending is quite different. Up to the point where the man is hitting a gong underground, it is the same. But the man who hears the sound of the gong digs by himself to find an old priest earnestly reciting a sutra. After the man takes him out, the two talk to their hearts' content under the moonlight. It is possible that Akinari could have produced such a form, close to a religious dialogue."

"But isn't this present form of the story more like Akinari?" I protested. Retaining the original form over several decades by having the figure recite the sutras after entering meditation as an expression of a fanatic faith might have been attractive to the argumentative Akinari. Yet it seemed to me that the latter half of the version which Professor Nunokawa was translating contained a far more deeply penetrating ghastliness and a more profound sadness.

"Ha, ha ha ha."

Sensing something, the professor laughed weakly, his protruding adam's apple trembling.

"Of course you too wish for a love union in the second life. It is all too natural—"

It was the professor's habit to make a not-so-graceful joke whenever he regained some strength. I crawled to the desk and said,

"We still have some time. If you are not too tired, shall we finish this tale?"

"Yes—shall we do that? Then it will be easier later."

Lying on his back, the professor again opened the book on his chest.

"Indeed the teaching of Buddha is simply untrustworthy. Over a hundred years have passed since he went underground and began hitting the gong. It is absurd that no trace of the soul was left and only the bones remained— We have come to his point, haven't we?"

"Yes, that's right."

Seeing the idiotic ways of the man they had uncovered, the mother of the master gradually changed her attitude.

"For all these years, I have been charitable and donated generously, besides never neglecting to recite the sutras. I did this only because I wished to escape the hard punishment of the next life after death. But seeing him I feel as if I have been bewitched by a fox or a badger."

Having said this to her son, she began to enjoy life, taking her daughter-in-law and grandchildren along with her to the fields and mountains to view flowers and the moon. She did so merrily, without worrying about the eyes of others, although she did not neglect to visit the graves of her parents and husband on the anniversaries of their deaths.

"Having good relations with my family, being kind to my servants and giving them things generously on occasion makes me live peacefully, forgetting that I felt grateful when I recited the sutras and listened to preaching." Mother often told this to people and behaved joyously, enlivened, as if released from confining bonds.

The dug-out man was always absentminded, but despite this, he easily became upset when he was not given enough to eat or when he was scolded, and would complain with a menacing glare. The servants and neighbors no longer treated him with the slightest respect. Even so, the name *Jōsuke* was given to him because he had entered *jō,* death through Zen meditation, and been resurrected, and with this name he worked as a servant at the house for five years.

At this time, there was a poor widow living in the village. The woman was considered dim-witted, but she became intimate with Jōsuke before the villagers realized it, and soon they began to notice Jōsuke plowing her narrow fields, as narrow as a cat's forehead, or washing her pots and pans in the stream in back. Since the master had kept him only because he was forced to, everyone suggested to him with a sardonic smile that Jōsuke became the husband of this woman when their relations became common knowledge. Finally he did become her husband.

"He says he does not know how old he is, but he obviously did not forget the ways between man and woman."

"Indeed there was a reason for Jōsuke to return to this world. I thought it was because his high spirit desired to attain Buddahood that he continued to hit the gong day and night in the hole, but the real reason must have been that he wished to return to this world because of his obsessive desire to be united with a woman once again. What a disagreeable wish." People talked to each other in this fashion.

Young villagers, curious about the way Jōsuke made love to the widow, went so far as to peep through the cracks in the wooden panels of the dilapidated house, but returned unsatisfied, for they did not see a ghost playing with a woman.

"It deprives me of my faith in cause and effect preached at the temple when I see a case like this with my own eyes." Talking like this, not only the villagers, but people in nearby villages as well began to neglect making donations to the temples.

A certain priest of an established temple in the village became deeply concerned about this.

Even though there is no way for an ordinary man in this degenerate world to know the far-reaching and flowing phase of the Buddha's saving methods, present events cannot be allowed to denigrate Buddha's virtues, he felt. He thought he must prevent men and women from straying by investigating the manner in which Jōsuke had entered the state of meditation. He checked the historical records and visited every old man in the neighborhood, trying to uncover the deeply buried truth of the matter, even to the point of neglecting his duties before the altar. Unfortunately, however, the village had been destroyed about a hundred and fifty years earlier by a flood; the houses and people had all been swept away. New branches of the river emerged after the flood, changing the geographical structure of the area, and with the improved water transportation, people came to settle there, gradually forming the present village. The village which existed before the flood now lay buried in the river. When it became clear that the present village of Kosobe was once an uninhabited sandbank, there was no way to find out how the coffin had come to be buried there.

"Yet when the floodwaters invaded, they might have flowed into the ears and mouth of the high priest and, drying hard there, changed him into an idiot like Jōsuke. Can this be possible?"

There were some who said so with dead seriousness and others who laughed with disdain, but Jōsuke's past was never revealed.

The mother of the mayor of this village was over eighty years old, and she became seriously ill. Sensing her death approaching, she summoned her doctor and said to him,

"This time I know this will be the end. Up to this time I did not know when I would die and have survived helped by the power of medicine. You have taken care of me for such a long time. Please care for my family after my death. My son is over sixty years old, yet he is so undependable that I worry about him. Please talk to him occasionally so that he will not cause the family fortune to decline."

Hearing this, her son the mayor grinned embarrassedly and said,

"I too have reached the age of gray hairs. Not being a bright man by nature, I'll never forget your teaching and shall devote myself to our family business. Therefore mother, do not worry and die peacefully reciting the sutras."

"Did you hear that, Doctor? You see he is such a fool. I do not wish to enter heaven by praying to Buddha; nor am I afraid of falling into the world of animals because of my lack of faith. At my age I have seen the animals of this world and realized that horses and cows do not always experience hardships as much as we are told. They enjoy themselves much of the time. Men are supposed to be far superior creatures in the ten worlds, but the times of happiness are few to recount; driven day after day, we have less leisure than horses and cows. All through the year we work constantly, washing clothes and dyeing them again and again. And when the year approaches its end, paying taxes is a matter of life and death. On top of our own worry, the tenants who are supposed to bring us rice come to complain about their own miseries—where and how can I ever reach heaven? But as my dying wish I ask you not to bury me underground in a coffin. Please bring me to the mountain and let the coffin be entirely consumed by fire. Doctor, please hear this as my witness. My last wish is not to be like that Jōsuke. Ah, now everything is troublesome! I don't even want to speak!"

Saying this she closed her eyes and soon passed away.

According to the will, the corpse was brought up to the mountain and cremated. Jōsuke, who had entered *jō,* joined the tenants and day laborers in carrying the coffin which was set on fire, consuming the corpse. Taking the role of a cremator, Jōsuke remained working until the bereaved family gathered the bones from the ashes, white and thin as twigs, and placed them in a cinerary urn. Knowing that he did so in order to receive as much as possible of the rice with black beans prepared for the deceased, people could not help but feel disgusted.

"Don't even dream of depending on Buddha to be reborn into heaven. Seeing that—"

They spat as they said this to their children.

"Yet Jōsuke was reborn and took a wife. It might be the will of Buddha to let him fulfill the promise of love which goes beyond life into the next life."

Some people spoke in this way, but the wife of Jōsuke, often fighting a husband-wife battle so typical that even a dog would know better than to interfere, would run into a neighbor's house, complaining and crying bitterly.

"Why have I taken such a good-for-nothing as my husband? I miss the time when I barely survived by picking up fallen grain. It would have been so nice if my former husband had been resurrected instead of him. If it had been my former husband, we would not have suffered as we do now from the lack of rice and wheat, and from the lack of cloth to cover our skin."

Indeed, many strange things occur in this world.

By the time the dictation was over, the short winter day had long since turned into darkness. The professor must indeed have been exhausted, and he lay with the book on his chest under the faint, yellowish electric light, closing his eyes repeatedly. No criticism or impression of the work, which might naturally have been expected of him when the translation was over, was forthcoming. Thinking of the trip back home which took over an hour, I myself felt restless, and without even quite properly saying good-bye, I hurriedly left the house.

Professor Nunokawa's house was located at the far end of the Nerima district where there are many trees which shed their red leaves in autumn. The area retained the feeling of the old Musashino section, creating an atmosphere in which those who came from the center of Tokyo cannot help feeling nostalgic. The house was off the main road where the bus runs, however, and it was quite hard, particularly in winter and summer, to get to the station; one had to walk a long way on a narrow farm road through woods and a bamboo grove. If I took the road in the opposite direction up to the main street, I could walk along a well-lighted road to the next station even though it took a bit longer, but I usually walked through the farmland since I was quite used to it.

Today, too, hurried by the radically shortened day, visibly shorter than it had been when I had come just a few days earlier, I plodded along the dark road, holding an umbrella for protection against a slight rain which had just started, my chin buried in the raised collar of my overcoat.

All the more because I had left without talking with Professor Nunokawa, the image of Jōsuke of "Love in Two Lives," which I had just been copying, floated in my mind as vividly as if he were a living man. In this tale, the story of Jōsuke's life before he entered meditation is never revealed. Is it because Akinari based the tale on some existing ancient story that he let Jōsuke fulfill his obsessive desire to love a woman, even letting him be reborn as a dull countryman to do so? Or was it Akinari's own intention? As Professor Nunokawa had pointed out, Akinari in *The Rain and the Moon* period would have treated the story differently. Then he would have narrated with ghostly beauty the story of a monk who, captured by the sight of a beautiful woman just prior to beginning his meditation, failed to enter nirvana. He would have created the story of a man whose obsession prevented him from breaking the eternal cycle of reincarnation, forcing him to continue to hit the gong. In contrast, Jōsuke in "Love in Two Lives" is quite slovenly and idiotic no matter how we look at him, so much so that he might well be a character in a comic story. It is likely that when he wrote this tale Akinari was half blind and had already lost his old wife Koren, who had become a nun in her later years. In his extreme poverty and loneliness, I conjectured, he must have written looking half-disdainful and half-frightened at his own sexual longing, still flickering unextinguished like a charcoal glowing under the ashes, a longing which stayed with him together with his unceasing desire to write. The idea that a virtuous priest who had reached enlightenment with regard to life and death was reincarnated to fulfill his worldly desire for love, a desire which he could not fulfill in the previous life, and that he had even borrowed the body of an illiterate idiot to do so, seemed to me to illuminate the eerie quality of his doting sexual longing, still wiggling like a worm. Akinari's skepticism about the Buddhist code of cause and effect, a skepticism expressed twice in the story in the form of the old women aspiring for salvation, may reflect the agony he felt over his unsublimated, cyclically returning sexual desire. On reflection, it struck me that the relation between Professor Nunokawa and Mineko was not entirely unlike that between Jōsuke and the widow. Professor Nunokawa must have attained Mineko, who was much younger than he, while she, knowing that he would not last long, had transferred the ownership of the house to her own name.

While thinking about these things, I recalled quite unexpectedly the night before my husband died from the bombing, when we made love for the last time. Not the memory but the feeling of my body when I wiggled like a playful puppy in his sturdy muscular arms and panted until reaching the final joy of the senses in which body and mind both become numb, as if they are about to be extinguished, returned to me. My womb throbbed. Startled, at that moment I slipped and staggered, almost hitting the ground with my knees.

"Be careful." As I heard the sound of a man's voice, I felt someone grasp the arm in which I was holding the umbrella, and I was helped to stand upright again.

"Thank you very much," I said, still breathing heavily.

"Around here the roots of the bamboo are often on the surface." After saying this unclearly, in a low voice, he asked,

"You didn't drop anything, did you?"

He stooped as if to look for himself. It was indeed in the midst of the bamboo grove, exactly midway between the professor's house and the station. The flickering light from the landlord's house behind the bamboo grove could be seen through the bamboo trees. In the darkness I could not see his face, but since he had no umbrella and his coat seemed to be wet, I extended my umbrella toward him, saying,

"Why don't you come under the umbrella?"

He then came close to me without any discretion, pressing his body against mine.

"It is very cold, all the more because of the rain." So saying he grasped my hand with an ice-cold, gloveless hand, pretending to help hold the umbrella. Although I could not see his face, his voice and figure were like those of a shabby middle-aged man. But the hand I felt through my gloves was as soft as a woman's hand. Since I like a man's hand to be bony with a strong grip, just as my husband's hand was, I did not like the softness of this man's hand. Yet strangely, I did not try to brush it off, and I felt a guilty pressure at the soft coldness of his palm steadily pressing on my hand through the gloves. Helping to support the umbrella with one arm, he placed his other arm around my shoulder, holding me tightly against him. My body was completely encircled by his arms in this way and I had to walk awkwardly, entangling my body with his. In the darkness I staggered many times, and each time he held me up again as if he were a puppeteer, each time touching my side, my breast, and other parts of my body. He smiled at these moments in an ambiguous way which might have indicated he was happy or sad.

I thought for a moment that he might be a madman, but the thought did not decrease the strange pleasure of being held by him.

"Do you know what I was thinking when I slipped?" I asked in a coquettish tone of voice as if drunk with wine. The man shook his head and embraced me still tighter, making it more difficult to walk.

"I was thinking of my late husband. My husband died in Kure in an air raid—while he was in an army bomb shelter. I was four or five blocks away in the army barracks, together with my son, and escaped the attack. I wondered whether he thought of me before he died. I don't know why but I want so much to know what he was thinking at the time of his death. He was in love with me, but since he was a soldier he used to think that to love me and to die alone were two different matters. I too used to think his way of life beautiful and admired him. But I wonder now whether he truly believed until his death that to love a woman and to die alone are not contradictory."

Without answering me, the man pressed his cold lips to mine, as if to contain my words by covering my mouth.

While his cold tongue was slimily entangled in my mouth, by chance his sharp cuspid tooth touched my tongue. It was, without mistake, that of my husband.

"It is you, ah it is indeed you—"

When I shouted, the man, as if asking me to forgive him, pushed me down among the bushes, hard against the raised bamboo roots, and covered me heavily with his body. Yet his hands were still soft and different from my husband's. Rejecting these heavy hands trying to unhook the buttons of my overcoat, I cried out weakly.

"No, it is not you. You are not my man, not him." Still without saying anything, the man held my hand and put my fingers into his mouth. Inside his cold mouth, the dog tooth was as sharp as a gimlet, just as painful as when it touched my tongue in the past. But the hand was different. My husband's hand was not as soft and fat as a woman's hand. And his body too— Then suddenly I recalled the smell of the moldy, sick man I encountered upon entering Professor Nunokawa's room.

Is this man Professor Nunokawa? The minute I thought this, my body jumped up fiercely like a wild dog, my voice shouting completely different words. "Jōsuke! This is Jōsuke! This man is—" Whispering this over to myself, I ran as fast as I could in the darkness.

When I stepped out finally into the well-lighted street in front of the station, my heart was still beating hard, still captured by the vivid fantasy that visited me on the dark road. A train had just come into the station. Through the narrow exit men in dark coats were being pushed out one after another from the flock of salaried men coming home as if they were being pushed out each in his turn from a mold. Standing at the side of the exit and watching the flock of these men being pushed out, I felt that every one of them was unmistakably a male. It made me, a woman, feel envious, and at the same time pressed my heart with sadness.

I felt surely that the Jōsuke of Zen meditation was living through these men. Even more than the shameful fantasy on the dark road, the thought made my blood throb, filling my heart with a sensation that was comforting as well as disturbing.

Ants Swarm

Kōno Taeko

For a moment Fumiko had the feeling that she could return to the deep, free world in which she had been until a moment before, but after all she could not go back to sleep any longer. Wanting to get up, she pushed away the light cover with one hand. But she could not shift her position further. She was concerned with the change her body might feel when she moved. Pressed by the warm darkness, she stayed in the same position, lying on her side, bending her body slightly like a bow, and keeping her legs together completely all the way to her toes.

Her husband, Matsuda, continued to breathe strongly in his sleep. The pendulum on the wall clock swayed busily but did not tell the time at all. Lying down in the same position as if forced to do so, she gradually came to feel a sense of pressure on her chest, as if her breathing was being controlled. After remaining immobile in this way out of laziness and hesitation, she finally raised the upper part of her body, supporting herself with one hand. She did not feel any change. Untying her tightly pressed legs and raising her knee, she moved one leg to a cold tatami mat. This time too there was no special feeling. She stepped on the tatami.

Opening the paper door, she found to her surprise that it was not midnight. In the quiet corridor, she could see the shapes of things. The clouded-glass door looked weakly whitened. Since Fumiko's sleep seemed to become deeper toward dawn, it was very rare for her to wake up at this time. Thinking that she must have been worried even in her

---

 Translation by Noriko Mizuta Lippit.

sleep, she touched the doorknob of the bathroom.

Fumiko's menstruation was already about a week late. This had seldom happened. From her girlhood, her menstruation had been very punctual, to the point of being amusing. Even the time it started, usually in the evening, was the same. Since her marriage to Matsuda, a man a year younger than she, its punctuality had never changed. The two had meant to avoid having a child and had had no trouble at all.

When it was three days overdue, she told Matsuda about it.

"Hmm" was Matsuda's first word.

"Just as I said—I felt sure that time," Fumiko said.

In their married life, they depended on nothing but the punctuality of Fumiko's menstruation to avoid having a child. Fumiko was working for an American law firm; she left the office promptly and had Saturdays off. Matsuda, on the other hand, was a journalist responsible for political affairs, and his work schedule was irregular. When he was busy there were days when he came home at dawn or did not return home for two days. He was sometimes busy for weeks at a stretch. When he could finally take a breather, it sometimes happened that Fumiko was in a danger period. When that happened, Fumiko felt concerned and had suggested taking some measures other than their usual one. But Matsuda never considered it. He said no and avoided the subject.

"Lions eat at one time and they are free from appetite the rest of the time. I am like them. I don't develop a lingering attachment," he added. Fumiko thought she could understand.

When Matsuda's caresses started, Fumiko's body immediately desired pain. And seeing her desire pain, Matsuda, as if he had been waiting for it, became excited. He tortured her hands and feet, eventually using some tools. Finally, without being aware of it, she would whisper in a hoarse voice, "You love me as much as this." Then responding to the still further heightened effect that was evoked, she would blurt out, demanding crazily yet begging, "Please forgive me, forgive me." Finally, Matsuda finished loving while torturing. It was not unusual for him to want to share the following night in the same way. But when they entered the unsafe period, Matsuda was unbelievably clean. During that time he sometimes made a few scars on Fumiko's body, but he never bothered her more than that. And she loved this aspect of him mixed with his passionate aspect.

One morning, about half a month ago, she was dragged back to bed when she was about to get up. When she reminded him of the unsafe period, he answered irritatedly that he knew. Fumiko frowned. Even on the safe days, Matsuda would awaken with the sound of many newspapers being placed above his pillow, and after absorbing himself in them, he would fall asleep again. The reason why Fumiko felt miserable may have been because she was used to his usual clean behavior; since it was very rare for him to act in that way, however, she could not refuse him. Seeing him getting excited hastily and unable to rid himself of his desire, she finally urged him on.

Yet Fumiko's unpleasant feeling remained. Although that fleeting feeling she had of not caring gave her a sense of release, the fear of pregnancy was always with her. She had to leave for the office soon. Besides, it was love-making which omitted pain, and all the more for that reason she could not be immersed in it.

Getting up and making herself ready quickly, Fumiko felt even worse. Turning back she saw Matsuda under the covers, only his head showing.

"I feel I conceived a child. It felt different from usual," she said to him deliberately.

"Is that right?" Matsuda answered from under the quilt. She smiled bitterly. But at the time she was struck by the feeling that her words were true. She stopped her hand while drawing up the zipper of her skirt. When she had gathered all her impressions, she felt sure that there was a special feeling, despite the fact that she had been unable to lose herself.

When the usual time for her menstruation approached, she became more restless. And finally when the day came and it did not start then or on the day after, she was annoyed and nodded to herself, yes it was true. It may be true that there cannot be a feeling of conception. But she thought she must have felt it as if by some presentiment. She thought it was not only because of her fear that she had felt sure at the time.

It was not that there was no tone of resentment in her telling him about the lateness of her menstruation and touching upon that day.

"What are we going to do?" she said.

"What shall we do? We're in trouble." So saying, Matsuda held his knees in his arms.

"Yes, we're truly in trouble."

"But it's funny to say I'm sorry between husband and wife." Fumiko felt that she was being scolded for her tone of voice.

"Yes, that time, I too felt as if I were under some evil influence," she said. "But we must think about it. The time is the problem."

They were supposed to go to the United States in July to study for a year. Matsuda had received a Fulbright fellowship and was to attend C University. Since Matsuda had applied early, she too was to enter the university. She had already completed the necessary entrance procedures. Since they had to attend the orientation program there before the start of school, they had only about a month before they were to leave for the United States. Fumiko, who had not experienced this before, did not know for certain, but she felt it would be impossible to take care of the fetus and return to her normal self within a month. She had heard that it would be very difficult to take care of it in the United States. Even if it were not, it would not be possible to do such a thing soon after they arrived.

Matsuda listened to Fumiko's words silently and finally said, "But it doesn't mean that it's certain."

"It's been three days. It's never happened before."

"Relax."

"Yes, but if I wasn't mistaken—"

"Yes, that's right." After thinking for a while, Matsuda, raising his face, said, "Can you give up studying in the U.S.?"

"If it proves difficult to take care of before we leave."

"Will you do that?"

"Although regretfully, I will stay."

Then Matsuda said, "Then you will have the child."

"Do you mean what you said?" It took some time before she asked that.

"Yes. It's not bad to have one."

"What! When did you start thinking like that?" Fumiko gazed at Matsuda. "That morning, was that intentional—?"

"No, it was not. Absolutely not."

"Yes, I think so too. I believe you. But since when have you been thinking—?"

Matsuda did not answer.

"It was because I told you about it. Right? Isn't that right?"

"Of course partly because of that—"

"Is that right—I understand." Sighing, she averted her eyes.

Getting married while both were working, they had naturally avoided having a child. But when they gradually realized that they clearly did not like children, they promised each other that they would not have a child.

"There are very few men who want to have children badly. The reason why they make them, after all, while declaring that they will not have any, is because they are begged by their wives. I know at least two couples among my friends. If it's okay with you, it's fine with me."

When Matsuda said so repeatedly, she felt no resistance and would even take the initiative in swearing that she would not have a child. Recently they had become so used to their plan that they did not have to remind themselves of it.

Fumiko loved to hear Matsuda talk of his childhood days. "When my grandmother was dying, I couldn't be found anywhere. I was up on the roof shooting pigeons with an air gun. People in the house did not hear the sound of the gun, for they were out of their minds in that situation." He recounted, "Now I am middle-sized, but I was very small when I was in elementary school. When I entered school, I was the second from the front. Line up, the teacher would say and there was only one in front of me. I was very sad. I hated line-up." Saying this, he would stand up and stretch his arms straight out in front of him as the children were taught to do when lining up. At these moments she was attracted to him very strongly. When he was anxious to take off her dress in bed, he would suddenly become obedient, and putting his head on her chest like a child, put the button ripped forcibly off her dress next to her pillow, saying, *"Hai"* [here]. She could not help bursting out laughing with happiness.

Yet Fumiko had never wanted to have a child. Just to think of bearing the child and having to raise it disgusted her. This time too, when her menstruation was late, she was fearful, held a grudge against Matsuda, and was concerned only with finding a solution; she never had any real hope. Therefore she found Matsuda's words extremely shocking. She felt betrayed and jealous when she thought that Matsuda, while she remained oblivious, had come to think of a child and experience a parental emotion which she would never have, despite the fact that they had promised each other so emphatically. She also thought of Matsuda's childlike character which used to please her so much. Then, together with the feeling of betrayal, she felt that he was not grown-up enough to have a child of his own, forgetting the fact that although a year younger than she, he was thirty years old. This made her even more irritated.

Despite all this, when she said "not yet" after laying the mattress as he told her to and returning from the bathroom, she was surprised by her own obedient tone. Besides, since the day when she told Matsuda about being overdue, she went to the bathroom more frequently. It was not much different at work, but at home she felt like going to the bathroom quite often. Coming back from the bathroom she always reported to Matsuda whenever he was at home. He had never asked, but she knew well his tension when resisting the desire to ask her. She reported to him tenderly.

This did not mean, however, that she had already made up her mind to have a child in case she was pregnant. In her mind, she was thinking that if it did not start within a few days, she would investigate ways to take care of it. If everything could be taken care of before the beginning of the term, then of course she would go abroad. In that case, even Matsuda would agree to give up the child. But what if it could not be taken care of before that time? Then she would have no choice but to remain here; even then she did not imagine that she could come to want to bear the child. Despite these thoughts, the way she reported to Matsuda became more tender each time. To Matsuda, it meant that he had almost obtained her consent.

Fumiko came to notice that Matsuda's face shone each time she told him that it had not yet started. Then he would say,

"The tiny one is there, isn't it? It must be like a dot now. Or is it like a sesame seed? The sesame child, we must take good care of it."

Influenced by his manner, the words which Fumiko used in reporting changed to "It's still all right."

"Are you happy?" she even added sometimes, unconsciously.

"Of course I'm happy," Matsuda answered.

"You want me to bear it," Fumiko said as if appealing to him, but Matsuda only nodded two or three times, burying his face in her chest When Matsuda made such a gesture, like a child expressing a heartfelt desire with constraint and worrying about her response, she felt that she could not win. Yet since it was covered up in playfulness, it was easier for her to give him a response which was different from what she really felt.

"Do you want me to have the child?" she said. She wanted to see Matsuda's childish gesture again. Each time Matsuda would nod while burying his head.

"All right, then I will bear him for you," she answered each time.

It was only a matter of days, but Matsuda's fantasies about the child became inflated. Because Fumiko's attitude was such, Matsuda did not hesitate to share them with her.

Seeing Matsuda like this, Fumiko could not but wonder. Her wish that her period would start had never completely vanished from her heart. It was not only because of their study abroad. It was because she did not want to have a child. It was because she thought of the helpless feeling of conceiving a child which she did not want, and having to bear and raise it. She was stunned when she thought of herself letting Matsuda dream of a child and then letting his dream immediately become so overly inflated.

In spite of this, Fumiko did not dislike hearing Matsuda talk about a child. She even encouraged him to do so, thereby compounding her crime.

"The child will have been born by this time next year," she said.

"That's right. But I feel really sorry that I can't be with you when you are in labor. I wish I could pull it out from between your fingers," Matsuda said.

"I'll be fine, you don't have to worry. By the time you return—"

"A little one will be waiting for me. It really will be nice. I will love him. I will buy many things for him."

"What kind of things?" Asking, she pulled out a cigarette. But noticing Matsuda's worried face, she put it down.

"Dolls, tricycles— Then I will take the child to the zoo. Is there a book of tickets for the zoo?"

"A book of tickets?"

"I will buy a book of tickets to take him to the zoo. When he grows up, I will arrange it so that he can charge all the drinks he wants anywhere. If he does not drink much I will be worried."

"Oh dear! You really are an indulgent father."

"Yes, indeed I am," said Matsuda. "I will attend PTA meetings."

"Yes, please do. It will save me the bother."

"I may have to bow as deeply as this when I meet his teachers." So saying, he bowed his head so deeply as to hit the dining table. "—But if a teacher complains about my child, I will beat him up." She could not stop laughing. While laughing she thought she would like to see Matsuda play a father's role. She did not want to bear a child. Nor did she have the broadmindedness to allow Matsuda to have a woman other than herself. Yet a desire to see Matsuda play father welled up in her strongly.

Fumiko knew that her period had started. First she felt relieved. Then she thought it had been late because her fear and tension had been so strong since that morning. She thought again that there was nothing to worry about for a while. The matter of going to study abroad came back to her as before.

But when coming out of the bathroom to return to the bedroom, she stopped suddenly, looking down the quiet corridor at dawn. She looked around. There was a wash basin at the side. On the shelf were two or three razor blades which Matsuda had used and left there. The door to the kitchen a few meters away was open and she could see a cupboard there. At the end of the corridor was the entrance to the house. At the entrance, a short, dark, Japanese-style curtain was hung; it was for winter use, but had been left hanging for a few months now. The platform where the stairs started, the paper door to the bedroom, and the door to the western room— More than a year had passed since the two had moved to this rented house. They had already become accustomed to the place. Yet their life there in the past week had been completely different from before. It was particularly so since Fumiko confessed to Matsuda about the delay of her period.

Fumiko reflected deeply on the past several days. At night Matsuda's footsteps had sounded different. While she was preparing evening snacks for him, he had been eager to talk about the child. Since Matsuda usually had fallen asleep again when she was ready to leave the house for the office, she was accustomed to saying good-bye to him casually, but it had been different these days. Whenever I opened and closed this door—thinking of it she looked back at the bathroom door. She realized that she could not dismiss this incident as a foolish experience. She realized that the familiar curtain, corridor, wooden door, wall, and pillar had all been imbued with a certain richness during these days, but that the richness was now diminishing rapidly.

As she brought her body into the bed in the pitch dark, Fumiko thought of the disappointment Matsuda would feel when he learned that it was a mistake, and thought she might keep it from him for a few days. Yet raising her head which was almost sinking into her pillow, she moved closer to Matsuda. His breathing in sleep stopped, and murmuring something while half-asleep, he responded at last as if only his arms were awake. His cheek was warm. She embraced his head again, pushing her nose strongly against it and inhaling the smell of his ears which she deeply loved.

"It was a mistake," she said away from his ear a little. Suddenly she started sobbing. Realizing that Matsuda had raised his body sharply, she said again weeping, "It was a mistake."

"Is that right?" Matsuda's voice showed disappointment. "But it's all right. It's all right. Wait a second." The light was turned on, making a "*pachin*" sound.

"Oh, don't turn the light on."

"All right, I'll turn it off." So saying, he pulled the cord once again.

"I should have kept silent from the start. I talked too early," Fumiko said on the pillow which Matsuda let her share and cried again. She felt that she was like a barren woman who had been longing for a child and who had to face the bitter fact once again. Yet the tears flowed spontaneously.

"No, it was nice that you told me. It was good that we could have a rehearsal," Matsuda said.

"You were so delighted."

"Why not truly make me happy next time. First we will go to America together and then it will be fine when we come back."

"I will do it before it is very late," Fumiko said, and it was a truthful voice. She felt that she would like to let Matsuda have his child. "Although I do not like children."

"You will change. I too did not like children before."

"What if I don't change? What if the child is born and still I don't like him?"

"It wouldn't matter. I'd change the diapers."

"You must be an indulgent father. But I'll be different. It will not be so extreme if it's a boy, but to a girl I will be very strict. Since you will be exceptionally sweet and I will be extremely strict, other people will think I am a stepmother." So saying, she felt a strong feeling toward her stepchild-like child arise in her heart. It was certain that the incident had affected her. Yet she was released from the worry of the past several days. For this reason, the feelings about the child they had experienced stimulated her imagination extravagantly.

"If it is a girl," Fumiko said, "I will not let her have much education."

"I agree," Matsuda answered. "School education will surely do no good."

"I might let her go through compulsory education."

"No, that compulsory education is the worst. Let's go with tutors."

"Such a luxury! I'll never agree to that," Fumiko said. "It is more than enough for her to go to a local school. After she graduates from junior high school I will make her stay at home and use her as a maid. Yes, by this time of day she'd be forced to get up to prepare breakfast. I'll be here with you like this."

"That will be good."

"Do you like it? Then I'll make her do it when she enters elementary school."

"Can a first-grader cook rice?"

"I'll make her do it."

"What a poor child our child will be."

"But it will be good for the child. It's for the child."

"That's true too."

"I hate a girl who thinks of all sorts of things. Let's raise her to be a girl who never talks back. It won't do if she just restrains herself from talking back. She ought to be a child who is incapable of criticism, a child who has no opinion of her own. A child who can automatically do what she is told—an idiot-like child—even her face must look like an idiot's."

"You mean a child like a doll."

"Yes, like a Chinese girl in olden times; we'll let her learn only domestic matters from an early age, and soon after she graduates from junior high school, I'll marry her off."

"I'll visit her often to see how she is. I'll bring her some of the sugar we receive without your knowing."

"No!"

"But you never cook with sugar. There's lots of unused sugar."

"Is that so? Okay then, you can bring it."

"I'll go to her every Sunday."

"You'll be disliked by your son-in-law."

"No matter. He'll be a nice husband. I'll find her a nice, gentle one."

"It won't work that way. I'll urge him to be a bad husband, to be dissipated. You too must encourage him to be dissipated. Even if you find beautiful women, I know you will give them up. Bring them to your daughter's husband, as many as possible, like you bring sugar."

Fumiko continued, "I will not listen even if she comes to me to complain. I'll show her my body and tell her that her father is such a cruel man, but that her mother endures it. I'll tell her that she must bear it too."

The clock on the pillar struck. It had stopped after one.

"What time is it?" Matsuda asked.

"It must be six-thirty."

"Go look. It must be seven-thirty. You'll be late."

"That's all right."

Fumiko did not want their delightful conversation to be interrupted further. She was eager to resume it as soon as possible. But Matsuda got up, and making a noise as he unlocked the glass door, opened the outside rain door slightly.

"It is six-thirty, as I thought," Fumiko said, looking at the clock from the pillow. Matsuda closed the glass doors where the outer wooden doors were opened and drew the curtain again, just as Fumiko had often done for Matsuda, who used to stay in bed until after breakfast. Then he walked out into the corridor.

Today Matsuda brought the newspapers in by himself, and sitting cross-legged on the tatami in his pajamas, started to open them one after another. Fumiko could not complain, considering his profession. Yet seeing after a while that his eyes were wandering, she called to him from the bed,

"There's nothing, is there? It's okay, isn't it?" It was all right. Matsuda left the newspapers and returned to the bed, asking,

"Do you have time—is it still all right?"

"Yes," Fumiko answered, but already more than twenty minutes had passed. She certainly had to get up now.

"It is nice."

"I am envious that you can go late." While saying this she remembered suddenly that she had run out of butter for breakfast. She said,

"Is it all right if we don't have butter this morning? There isn't any left. I forgot to buy some."

"We have jam, don't we?"

"Yes, and cheese too."

"Then that's enough," said Matsuda. Through the opening left where Matsuda had failed to close the door completely, Fumiko caught a glimpse of the pleats of a skirt fluttering. It was the daughter she and Matsuda had created, a daughter whom strangers might think a stepdaughter.

"If our child forgets to buy butter, I will never forgive her," Fumiko said to Matsuda. "I am very strict with girls. Very strict. You will not be able to take it. Although you can do anything to me. You may hide in a closet and hold your ears with your hands."

The paper door opened. A short, pleated skirt came in. Fumiko pretended to be asleep.

"Mother," said her daughter, who was around seven or eight. Making the pause as long as possible, Fumiko said with her eyes closed, "I did not hear the sound of the door closing. You were scolded before for not closing the door tightly."

"But father does the same," the daughter said.

"Are you the same as father?" So saying, she jumped up and attacked the daughter.

"Do you mean to talk back?" With all the strength in her fingers, Fumiko pinched both sides of her daughter's mouth. The daughter tried to escape, but since she was pinched tightly from the right and left she couldn't move.

"Please forgive me, forgive me," she begged. They were the same words which Fumiko uttered at night in her frenzy when begging Matsuda for more pain.

"I cannot forgive you," Fumiko said. "It's not only talking back. Think of all the bad things you have done. To talk to your mother standing—who taught you such a thing? Also, when you came in here, you came in without saying anything. You must say something before you come in."

At the same time, Fumiko's fingers continued to torture her. Not only the short, pleated skirt, but also her blouse and undergarments, she took off everything and pinched her all over her body. Still she could torture her further about the butter. When she realized that, Fumiko brought the daughter to the kitchen, still torturing her.

"Is there butter?"

"We ran out. I was going to buy some, so in order to receive money—"

"You forgot to buy butter."

"Yes." As soon as she answered she crouched beside the gas range as if hiding and lowered her face. There were several red scars on her back. They looked like scars left by cigarette burns.

"Go and buy butter," Fumiko said. Butter in a yellow paper box slid down from the ceiling and onto her palm, feeling cold and heavy. Fumiko peeled off the thin, unwrinkled paper attached to it and scooping some out with a spoon, brought it to the gas flame. If it melts completely it will be very hot. Looking down at the back of the daughter who was squatting at her feet, Fumiko gazed at the butter in the spoon. A yellow lump melting smoothly from its edges—

"You'll be late," Matsuda said. The hands of the clock had already passed seven and were hanging down.

"Yes, I'll be late," Fumiko said, as if it concerned someone else.

"You must have been tired out. Because of this, since that time," Matsuda said. "Will you stay home? I'll call the office for you."

"Since I have never stayed away from the office—" Fumiko showed her decision to stay ambiguously.

"It's all right, now and then." Answering so, Matsuda's hands groped for her breasts. Being excited herself, Fumiko could not, like the other morning, think him despicable.

"When should I have a baby?"

"After we return from America."

"When after?"

"Any time."

"Isn't earlier better for you? Really, I would like to let you have a baby. But—" Fumiko felt Matsuda's fingers holding one nipple tightly. It was being pulled inward and the little finger of the same hand groped for the other nipple. The neck of the little finger missed the nipple two or three times. When it was finally caught, she felt severe pain in both nipples, as if they were being torn out. Yet Matsuda pulled them harder, with all his might, as if raising a heavy bucket. As they were pulled she gasped with pain and pushed her face against his shoulder, but this withdrew her chest further, increasing the pain and the pleasure.

"I would like very much to bear a baby for you, but as you know I don't like babies," she said as soon as she was able to breathe. "So please order me. Any time will be fine. If not, I will not—" Just then, her breath was cut short. Trembling all over her body with the pleasure of the pain and scarcely able to breathe, she said,

"Please, order me—please force me."

"I will. At the right time I will order you. 'Bear a baby even if you die,' " Matsuda said, strengthening his grip.

"I will bear a baby even if I die. So when I give birth to the baby stay with me darling," she continued to say, panting.

"Of course I'll be with you."

"To be with me is not enough. I may struggle because of the pain. You must tie me. That may not be enough. If not, will you hit me? It will be so wonderful to bear a child while being tied and hit. If it is such a childbirth, I would like to experience it soon— Please be quick."

Matsuda jumped up. Fumiko realized a strip of light was reflected on the curtain. But when she heard Matsuda open the closet and take out the fishing rods which were placed there although neither of them ever fished, she could not care about people knowing.

At the doorway, Matsuda, who had already put on his shoes, received the bottle of milk from Fumiko and brought it to his mouth.

"Are you all right?" he asked, taking the bottle from his lips a little. Fumiko smiled silently.

"When you become pregnant we cannot do what we did this morning for a while. I'll have to treat you with discretion. Be prepared for it, all right?"

"But when you can do it for me again, there'll be a child."

"Don't worry." So saying, Matsuda emptied the bottle, pushing his head back. "I'll make a soundproof room. Even now we need one. If it's going to be like this morning—" Handing the milk bottle to Fumiko, Matsuda said,

"I'll call your office." He left.

The bed was not yet made. The outer rain doors were all open. There was a little pause in the rainy season and the early summer sun was shining and the wind was blowing.

Fumiko sat down near the open corridor. She indulged herself in the sense of her body. Everywhere on her body, the hotness turned to pain and the pain turned to hotness while both gradually diminished. She liked the feeling—as if her body were rippling under the touch of the early summer breeze. Since they had used up their time and Matsuda had left without breakfast, she did not have breakfast. It was not that she was not hungry, but she continued to sit in this way, and gradually she became sleepy. She lay down in her clothes on the unmade bed and closed her eyes. She felt again that it was a pleasant, early-summer breeze.

When she realized what time it was, it was already close to two o'clock. I cannot let my child see me this way, she suddenly felt. She made the bed and went to the kitchen.

When she was about to open the window, she saw something jet black about the size of her palm on the bay window. It was wriggling. For a moment she could not tell what it was, but on closer examination it turned out to be a chunk of beef on which numerous ants were swarming fiercely. It was in fact lying on the wooden board. It was the meat which Fumiko had earlier put in the nearby refrigerator.

Matsuda had placed one of the pieces before they were infested with the ants on one of Fumiko's scars this morning. Fumiko had had him place them there several times before. Matsuda used to bring the pieces of meat with chopsticks. Seeing them brought dangling to her hip or shoulder, she used to burst out laughing. This morning Matsuda must have forgotten to put the wooden board holding the meat back into the refrigerator.

Numerous ants completely covered the meat. They all swarmed on it eagerly. There were a few ants which were wandering around the wooden board. And strangely, since all the ants had come to the meat, there was no line of ants leading up to it.

Nothing could be done with the meat now, even if she got upset. Since there was no need to chase the ants away quickly, Fumiko looked at them swarming and wriggling as if they were one body. She had not realized that so many ants lived in her house. She had never even seen them. Was it because her family had little to do with sugar? The reason why this family had little to do with sugar was partly because Fumiko was not a domestic woman. Besides the fact that neither husband nor wife liked coffee, and therefore did not need sugar for that, she had never tried to have a baby who would drink milk or a child who would throw away candy wrappers. She herself preferred smoking to eating snacks and she rarely spent time preparing dishes using sugar.

Yet there were so many ants living even in such a house. They must be ants which had forgotten the taste of sugar, or ants which did not know the taste of sugar. There must have been a taste of sweetness in the blood of the beef which, not having been put back in the refrigerator, had become a little sour. It was as might be expected of ants that they did not miss the meat, and came to swarm in such huge numbers.

Fumiko tried to observe the movement of individual ants. They appeared to be trembling rapidly and to be rubbing their heads on the meat eagerly. After a while, however, she could not continue to look at them since there were too many ants swarming too busily, too close to each other. As Fumiko watched them, the ants came to appear to be one black, wriggling thing. The black lump of ants continued to wriggle as if laughing at her, as if encouraging her.

九

To Stab

Uno Chiyo

I

As one incident was settled and things were again restored to their usual calm, my mind became filled with a restlessness unknown to others—how strange it was.

I didn't like this feeling. If there were anything to replace it, I would like to make an exchange—this was all I thought about. Would that I could escape from this feeling! I thought of everything that seemed within my power. And I even unconsciously planned something that might hurt others if carried out, simply because I wanted salvation for myself. What I will narrate here is a series of those events.

"I think now's a good time to publish another new magazine," I told my husband one day. Our conversation took a surprisingly uninhibited turn when it came to the matter of a job like this. "New? Publish another, different magazine?" "Yes, an entirely different kind. A small format, a decidely popular variety." My husband looked at me in silence. It is not necessarily all that easy to achieve a popular character in planning a magazine. However, at that point I wanted to completely immerse myself in the challenge to advocate the unreasonable: I was bracing myself with a feeling too unnatural to be called ambition.

"I'm thinking about having Nakata edit it. He's mature. He won't be controlled by his feelings; we needn't worry," I said, one plan after another for the magazine already starting to take shape in my mind.

Sasu (1963–66). Translated by Kyoko Iriye Selden with the permission of the author.

Later, reminiscing again and again about that time, I could not help feeling an indescribable chill, as I came to know what drove that dogmatic plan into being.

Whatever the motive, the plan, once conceived, seemed to stride forward under its own power as though it were a living creature. My husband did not agree with my proposal, yet he dared not oppose it. Is that because that would have required a different kind of determination, entirely unrelated to running a magazine? I irrationally interpreted his noncommittal attitude as consent.

To think of it now, nothing was so strange as my psychology then. It was as if I wished to pour into this new project the same passion as my husband did into his affair, or rather, as if I wished to spend on this new project the passion of a foolish wife who tried to stir her husband's feelings by also having a new lover so as to rival his affair. I started to make preparations with a sort of crazed, inordinate interest.

I went out every day. I invited someone to a coffee shop in the city after dark to talk about editing plans, or searched warehouses in the dusty back streets in order to buy paper through routes other than legitimate ones. While doing that, there were moments when I forgot why I was doing such a thing, moments when I was oblivious of what my purpose was.

"Listen, I beg of you. I mean to put my heart and soul into this job. I mean to try doing it to the absolute limit of my power." While sitting face to face with the man to consult about editing, it occurred to me that my inordinate zeal annoyed him. "I'm sorry. I have no intention of making you responsible. I only want your opinion for my own information." Placing a piece of paper on the table, the man explained his plans in a gentle tone that might suggest a certain lack of enthusiasm. While watching his large though for a man delicate hands, I involuntarily started to adopt a somewhat impatient tone as though I were wooing a man I loved. In that tone, I told him how I depended on his thoughts and also how much expectation I had for them.

I was imagining that if anything could set free the feeling that imprisoned and depressed me, I wanted to cling to it, throwing away everything else. I didn't forget my purpose. And, in fact, in moments when I was absorbed in my job, I forgot to think about where my husband might be.

The work seemed to proceed smoothly irrespective of my worries. The atmosphere inside the company, as the work advanced, suggested an understanding: Since it was urgent to bring it to success anyway, why argue about whose thoughtless plan had started it? Huge sums of money were poured into it.

II

The new magazine was out. It was a small, inconspicuous magazine. Yet, it was a popular magazine, which would be read only if conspicuous. I have no recollection of going to see how copies of that inconspicuous popular magazine were displayed near the front of bookstores. In view of my inordinate interest since the beginning, this was strange.

The world had changed. Our first magazine, the day it was published, had attracted people queuing up the narrow staircase to the fourth floor of the building. The line continued outside, winding around the building, and down the little side street. There was even someone who brought us vegetables, explaining that they were from his own garden. But the situation had changed completely. Two years since then, the world had become quieter, and many magazines were being published. People selected magazines. It didn't have to be ours. Other magazines would do just as well. Even if a new magazine was published from a novel project and with outstanding content, it took time before it became known to the world. Did I have this in mind?

In two months we discovered that sixty percent of the copies were being returned. However, publication was not stopped. Fortunately or unfortunately, we thought that the company's accumulated profits were still substantial. This dangerous project, therefore, didn't seem to merit so special an expression as "waging the company's fortune on it." We thought we still had room to experiment. And the great damage would easily be healed in time, bringing more than enough profit to cover the loss. There was no need to calculate how much surplus the company had at present and how long we would have to wait until the wound was healed.

"It's better not to do the impossible. There's a limit to what our company can do." One day I heard my husband say this in the second-floor editing room. I was changing in the sitting room downstairs. Family affairs and company business were always mingled in our life. From the wide second-floor room, voices carried as though through a pipe. My husband was interpreting the word "popular," introduced when planning the new magazine, while also trying to modify the main point of the project. Hearing his voice then, I thought that he, too, had now joined the work. I still think that I felt a sort of peace in thinking this way. Forgetting why I had dared to undertake this dangerous task, I assumed that he, too, probably shared my worries. Was it because of such wifelike feelings that I felt an indescribable, pitiable sense of relief?

Peaceful days continued. Nobody in the company worried that this project might cause the decline of its fortune. As before, when the sun set, we often went out to a dance hall or a drinking place in town. Even if I was with my husband then, I no longer had the need to doubt whether the young women who made their appearance there were the type that might captivate his heart. For his lover was elsewhere.

It was at the end of that year. This was right after the war, and never before had the festival called Christmas stirred people so much in an atmosphere that resembled madness. I, too, wanted to forget myself, pretending to be involved in this merrymaking. I made various preparations with exactly the same kind of secret wishes as when I had thought of that reckless magazine project. When I saw myself in a mirror in a downtown women's clothing store, wrapped in a gorgeous evening dress made of imported material, what devil's prank it was I don't know, but the image, I still remember, was of a pitiable woman in costume, almost like a moment's illusion in contrast with my husband's lover whom I had never seen.

It was late afternoon that day. I was dressing amid the Christmas bustle that filled the two-story, white-painted wooden house. "Are you there?" called a slightly drunk man from outside the living room. Since the house was built without a threshold, there were times when we didn't even know if a stranger entered or exited over the dirt floor. Thinking of how I appeared in the eyes of the slightly drunk man who knew his way around the place, I remember standing there for a second as though testing myself in a way. Whether I wasn't trying to see how a woman forgotten by her husband could still look in another man's eyes, at this point I can no longer say. The man looked stiff. "Are you on your way out?" he said, looking up at me. In his unfocused gaze, I felt the flash of something long forgotten, and I still can't forget the joy which ran through my heart so unexpectedly. "I have a date today. I'm sorry," I answered. My unnatural distance then, I think, spurred my hidden pride all the more. The man left right away. His back was an exact reflection of my own downcast heart, but I wouldn't have realized it then even in a dream.

III

This old story is a heartfelt parable that always helps me laugh at myself.

It is about a scorpion and a turtle. Once when the turtle was going out to sea, the scorpion said, "Take me on your back, for I can't swim." The turtle declined. "You're going to stab me once we're out at sea." "What are you talking about? If I stabbed you, I'd sink into the sea with you, wouldn't I? I'd sink and die, wouldn't I?" he said. The turtle was finally persuaded and went out to sea with the scorpion on his back. However, the promise wasn't kept. When they were in the middle of the ocean, the scorpion, after all, stabbed the turtle with his pincers through the shell piercing the belly. "O you stabbed me, you're going to drown with me," the turtle said. The scorpion answered in a sad voice: "I know. But to stab is my nature. I cannot but stab. Bear with me, please."

Despite small progress in sales, within half a year or so it became clear to everyone that the new magazine was a losing proposition. However, nobody said that we should stop publishing it. The more they wished to stop it, the more difficult it was to say so openly. And although no one was unaware of the cheerful air containing something ultimately dangerous, something which vaguely permeated the company, everyone pretended not to notice it. I think it was because they feared that admitting it would mean criticizing me—who had gone ahead with the publication. A heavy, dark feeling enveloped me. It was not just that this project had ended in failure. I was forced to realize the folly of launching a reckless project in an effort to forget my husband's affair. However, the sacrifice was growing bigger. There was no one except me who could suggest giving it up. One day, I suggested it with determination. It was a relief to think that the danger was avoided, and I had never been praised so much as then. Even outsiders commented: "It takes courage to give up a magazine of this stature. Well done."

My error thus passed uncriticized. Although the numbers must have been unmistakable in the books, no one discussed it. It's not that no one saw through my miscalculations. Unexpressed pity toward me perhaps closed everyone's mouth, I now suspect.

After that event, calmer days resumed. However, unlike a windless day, the calm brought no peace to my heart. I don't understand what it was. There was a certain feeling, like a desire to throw myself into the wind, not satisfied, while sitting inside the house, just hearing the wind. The fable of the scorpion and the turtle came back to me.

A long time ago, there was a piece of land for sale on the east side of Ginza. The war had ended and such lots were everywhere. We bought it for no special purpose, and at one point used a zinc-walled barrack we had built there as a storeroom for such things as paper and returned magazines. It was a mere whim in those days when there was no way to use the money we had, but interestingly none of us thought of putting the land to more meaningful use. Even when the world settled and the real value of such land was gradually recognized, no one mentioned the lot. It can be said that our company was still affluent enough so that we didn't have to think about the fact that we had such a lot. One evening, on the way home from a walk, I passed by the lot, which was surrounded by a plank fence. Through the planking there was just one lighted spot. I could vaguely see tall grass bending in the wind. A thought crossed my mind: why not build a house to live in? Sudden as it was, the idea captivated me—it was a strong feeling that could not be called just a whim.

To desire a house-like house to live in, just the two of us together, would seem normal. However, what was in my mind then wasn't normal. It was something that I had never even thought of before—something that lurked in my consciousness.

My husband often went to our other house in Atami then as before. Maybe he did so more frequently than ever. However, as before, I didn't know whether he really went there to work, or stayed at an inn somewhere with his lover. No. I didn't try to find out. Yet, although I pretended that I thought he was really working at the Atami house during every stay in Atami, I experienced unrest in my heart which I could not express to others. O to escape from this unrest—I prayed. I knew, however, that this wish was unfulfillable. If there were a house in Tokyo, and right near the company, much more spacious and more comfortable to live in than the Atami one, maybe my husband would go to Atami less often. Or rather, by depriving my husband of the excuse that he was going to Atami, I might be able to reduce the frequency of his meetings with his lover. Did I perhaps think this way? How foolish this childish conjecture was, I later realized.

That very night, I told him about this plan. Once put into words, it was as good as started. I said, in a funny, cheerful voice, "Why have we left that place alone like that? Isn't it just the right kind of lot? So near the Kabuki Theater, and yet as quiet as if nestled in the mountains. Listen, you'll be able to do your work in Tokyo; besides, don't you think it's about time we built our real house?"

My husband didn't answer right away. His face revealed no joy concerning the prospect of having our own comfortable house in Tokyo, and right near the office at that. However, this was also his habit at any given moment. Between us it was customary that whatever was directly related to living was for me to deal with and carry out at my discretion. Now that I think of it, this was a strange habit. My husband was never uninterested in matters of living. Yet, his joy came later when he had become used to a way of living. He never expressed joy instantly. This time, too, I thought it reflected his usual habit, so I never really thought about his feelings.

IV

It was when the magazine, whose publication had been discontinued, was finally forgotten. The empty lot was prepared; the barrack was demolished. The cement foundation was laid, and framing was erected.

Someone in a responsible position took charge of the design, so it was expected that great care would go into the construction. The best materials for those days were collected. This, too, was simply what my hidden thoughts brought to fruition, I now realize.

Over and over again I went to the construction site. Whatever the reason, building one's own home should be a matter of special concern. However, each time a room took shape—now the living room, now the bedroom—was there never a moment when an empty wind passed through my heart? Once I invited my husband to go to see the site with me. It was a fair fall day. In his light-colored coat, he walked lightly as if flying on the wooden planks, talking about this and that to the men at work. In spite of his enthusiastic tone, I perceived that his heart wasn't there. Yet, thinking of myself coming to see the construction site with my husband on this fine fall day, I could not think of it as a momentarily fabricated peace. Nor did I even remember that it was a certain scheme that I had created on a whim. I cannot forget the high pitch of a wood plane resounding in the quiet city where no other noise was heard.

I didn't know how much money the company paid to build this house. Even when I heard that no one else spent as much on materials for a house, I never dreamed that we had to calculate the loss the company faced by that sum. In my heart, which desired to dare to build this kind of house, there was not a bit of the usual satisfaction of obtaining something priceless. There was a vacant feeling as though it were someone else's affair—strange on reflection from this distance.

When the house was completed, I don't even remember. Our colleagues merrily carried books and what not, I think, through the clear city, but perhaps I remember it wrong. Our luggage was that little. In that house, not the old furniture that we had used until then but specially designed pieces were placed in each room. Despite the fact that they were done in quiet, genteel taste, the rooms seemed somehow remote within, as though the house were another's. It was like coming into a place that wasn't really ours yet.

"I see bamboo grass from my room." I turned back to my husband. In the yard of the small house in the suburbs where we had lived till a while ago, too, the same bamboo grass was planted. This small resemblance put me at ease, yet the joy of living with my husband now, starting today, in this house, did not swell at once. If my feeling in those days really was to seek peace for my husband and my only wish was to live in the same house together, I would have been overjoyed now that such a fine house was built and that we were going to live there starting today. I don't understand. Perhaps it was a foreboding that my fate was to be trapped into leaving this place soon after settling in.

V

The town abounded in big restaurants and geisha houses. Even during the day there was the sound of *shamisen.* Since it was quiet thereabouts, sometimes a guest's voice was clearly heard just outside some window. Such odd things were at first interesting, but soon they became part of our daily life. That I had my house built amid the gay quarters and hadn't minded it may perhaps have indicated our unusual psychology in those days. It was probably about two hundred yards from our new house to the office. It became our custom to walk to work separately, each at a different hour, through town. At times we walked together, looking at the display windows of the same shops, talking about this or that. Both my husband and I worked in our separate studies, and when we finished, we went to the office. Having the house and the office in different places seemed to punctuate our life—but only for a while. Something unexpected, or rather, something that was clearly anticipated, occurred. One evening, my husband didn't come home, without having told me where he was going. The next day, and the following day again, he didn't come home.

I discovered then for the first time how ingeniously this new house was built for waiting alone for my husband's return. My study was on the east side on the second floor. Adjacent to that was the Western-style bedroom. As long as it was locked, no one opened my room all day. No matter how I threw myself in anguish in the spacious, hence all the more vacant, room, no one saw it. No one even knew whether I was there. However, although this might have been difficult for others to imagine, I knew what kind of emotion jealousy was, having embraced it in my heart for so long. It was something that I had been familiar with for such a long time and, so to speak, something that I knew well. If I said that this familiar feeling was a sweet one, would anyone believe me? Even to cry was, if such an expression may be allowed, pleasant to me.

The bedroom walls had built-in bookshelves. As I was randomly pulling books out and looking at them, something fluttered to the floor. It was a small photograph.

I don't know how to explain my feeling then. It was a picture of an ordinary young woman without any particular distinction; yet at first sight, strangely, I thought it a picture of my husband's lover. Someone had certainly told me that she had long, showy hair, dyed red and swept back. But the moment I saw the face, almost smiling and somewhat coquettish, and the Japanese-style hairdo, I knew unmistakably it was she. Yet at the same time, what a change in emotion took place: as I realized that this was the person who had till this very moment been the object of my affliction, of my jealousy, to such an extent that I had feared to learn even the color of her skin and hair, her face that seemed to begin smiling arrested my heart with its almost childlike tenderness. It even suggested the color of her soft skin and seemed to penetrate the human heart. My mind became calm. If I try to describe it, I had absolutely no feeling that I had been robbed of my husband by this person; it was as if I could even accept it, should she rob me of him. If I say that I understood his affair at that moment, would anyone believe me? I felt that I understood, as if it were my own affair, why my husband neglected me and went to this person. However, it was all right with me. I had wanted to vie with her for one husband until that minute. Where that feeling had vanished or why, I didn't know.

VI

It was in the afternoon three days later that my husband returned. Is a woman with a job lucky on such an occasion? As if nothing had happened, I climbed up the stairs to the office. The bath near the entrance and the living room inside, where we had lived until a while ago, were both being refurbished. That was because I had suggested remodeling them into some kind of shop, since we were located in the middle of the Ginza. "Make the display window big. It's fine if the whole store looks like one big window," I told the men working there, and went up the stairs. Then I saw my husband doing something with our colleagues by the side of the editors' desk which had its back toward me.

My husband turned around. He said "Ah," not quite a word, and smiled a little. I knew well the way he started to talk with that smile. And even though it might have been just a moment's gentleness, I liked that smiling face. I wanted to tell him, at once if possible, how I had felt while waiting for him, right in front of all the office people looking at us. My feeling had not changed. I was not critical of his having stayed out for a number of nights. No, I didn't resent the three days that he probably spent with his lover. A strange thought had seized me. I had stepped down of my own accord from that seat of competition where I had fought for him against his lover.

This may be difficult for others to believe, yet I understood very well in my heart the road I had taken to that point. I remembered the time, long ago, when we first talked about getting together. We ignored, since we were younger than now, the age difference between the two of us, one that was rare for ordinary couples. Well, I can't say that it didn't concern me at all deep down, but I wasn't fond of thinking of the remote future in relation to that. Between us, with our unusual relationship, it was thought natural that we lived the present moment to the hilt, instead of worrying about what was to come. And, time having elapsed, those future things which we should have worried about earlier were now approaching before our eyes. What could I complain about, and to whom? Owing to my long habit, I told myself on such an occasion, "This is what I have done myself." I had learned how to put my thoughts in order in my own mind. And this seemed natural even if it were, in anyone's eyes, my last experience as a woman, and moreover the most painful.

"We can't use this photograph. But I wonder if there's another one instead," my husband spoke, not to me, nor to anyone in particular. It was the last day for selecting a photo for the frontispiece. I went to his side without awkwardness. There was the faint smell of the soap my husband was accustomed to using. In the same way as on any afternoon when nothing had happened, we were working together for the company.

Suddenly, an event occurred which seemed to erase in one sweep such things as our marital problems. Maybe it was fortunate for our relationship that it took place exactly at that moment. That was how I sometimes felt later. I heard the noise of many feet downstairs. Following loud voices as if arguing about something, the disorderly footsteps came up to the second floor. They proved to be fourteen or fifteen people whose arrogant air indicated that, although not policemen, they had a certain authority. It happened in a flash of a second. As they stormed into the accounting room in the farthest corner of the second floor, they swept together, so to speak, and crammed into their briefcases everything down to a torn scrap of a memo pad, not to speak of every single account book. They even rummaged through and examined the contents of the handbags of young women who didn't know where to run and of the trash box in front of the office. "Nobody go outside," one of them said in a loud voice. And they also took small notebooks from the pockets of the members of the company. They searched every place in the office for as long as two or three hours, and then they went off.

It was after dark when we comprehended what this was all about. "For the time being, I think it's advisable that you directors stay away from your residence," said someone in the sales section. Probably he feared that since we were not so knowledgeable in matters of finance, we might respond to cross-examination to the detriment of the company. Strange as it may sound, we had no knowledge of how the company's finances were managed, or should be managed. Rather, it seemed that it was done so smoothly and accurately that there was no need to know. In other words, we were prepared neither to judge what kind of disaster it was, nor to plan how to break through the disaster, in the event of such a sudden accident.

After dark, we left the office with other members of the company. When we gathered and talked together at a certain hotel in Yotsuya where we had arranged to meet by telephone, for the first time an unspeakable terror touched our minds, too.

VII

It was only after two or three months that the entire significance of this incident became clear to us. Needless to say, we had been working calmly standing on a precarious bridge. As a result of a letter, the company's finances had been investigated. Proof after proof was presented. A vast tax swindle was suspected, and we were subjected to a big roundup. It was then that I recalled the face of the person who had said, "Your way of profiting doesn't last long. Can't last long. When the world is settled and returns to normalcy, the time will surely come when you people will be in trouble." It can be said that we had been heedlessly doing, unawares, the kinds of things that would provide evidence of a vast tax swindle. How did that huge sum flow so as to attract people's attention? It seemed to me that the majority of that useless loss had been my responsibility, and piercing agony added to the sense of terror.

When one is placed in such a situation, strangely, one's mind becomes serene. My husband and I remained alone in a spacious hotel room. We were to stay there for two or three days, or even longer. It was strange to stay like that in an unfamiliar place, when our house was right nearby. "Well, why don't I take a quick bath," my husband said casually, standing up to go to the bath connected to the adjacent room. I heard a splash as he poured water over himself. That was a long time ago. Several times I had stayed, just like now, alone with my husband in a similar inn. I don't consider that my husband and I staying in this place now resembled the couple we had been before. When we were here as though cornered in this place, how could there still have been any hint of romance?

We lay in the beds prepared there side by side. In the unlighted room, sliding paper doors looked dimly white facing the verandah. What was shaking our minds in common was the incident of that day. Between waves of terror, an emotion ran through my heart. In the fact of sleeping in a bed side by side with my husband in an unfamiliar inn, there was a sentiment remote from terror, and resembling joy. The events of these three days returned to my mind slowly. Yes, now I would tell him how I had felt last night. Nothing seemed a greater joy to me than being able to tell him about it. It was a strange moment. "If this thing doesn't work out all right, I think it's fine with me if we quit the company. What do you think?" my husband said unexpectedly. I couldn't answer right away. Caught up in an incident which might lead to the collapse of the company, what was I trying to say? The thoughts that passed through his mind were transmitted to me. Terror had drawn our bodies toward each other as though we were two animals. "Even though you may not hold me, please lie side by side with me like this from time to time." This was all I wanted to tell him then. For the first time, tears ran down my cheeks. This seemed to me my valediction to something.

Facing the Hills They Stand

Tomioka Taeko

I

Fine to say he had come from the province of Yamato; but no one knew where that province was. The province of Yamato, like the province of Aki or of Hitachi, to people here was just some faraway place. What was more, the man from that province of Yamato did not know himself if there was in fact such a province. Anyway, a man came here from somewhere.

When the man came, there were only several dozen houses in the village. The houses stood here and there along the big river, and the rest was just reeds. This big river was near the sea and branched off into many small streams; thus this place was, so to speak, a village made up of many islets.

The man was no longer young, but there still was some time before he reached age thirty. He was trying to kill that time busily spending his days in a strange place. When he came to this village, right away he copied what the village women did: he skipped into a little river, scooped up baby clams, and ran to a neighboring village to sell them. Since he was swifter than the village women, selling baby clams instantly became his livelihood for a time. While thus occupied, he saw a heap of dirty hemp bags piled on a boat going up and down the little river. The boat

---

Oka ni mukatte hito wa narabu (1970). Translated by Kyoko Iriye Selden with the permission of the author. This story has benefited from the translation assistance of Hiroaki Sato.

stopped near a wooden warehouse on the river, and the bags were dumped into it.

Sitting by the riverside, every day he watched the hemp bags being unloaded from the boat. The dirty, tattered bundles just disappeared into the warehouse; the man did not know what happened to them afterward. Every day a boat docked at the warehouse, hemp bags were unloaded, and they disappeared into the warehouse; but the rest he could not figure out. Young men of that village, shouldering hemp bags, deftly crossed a plank laid across from the boat to the warehouse. What happened to the hemp bags stuffed into the warehouse—this was what the man wanted to know. To be sure they must be carried out somewhere before anyone noticed. Otherwise, that many hemp bags could not be stuffed in every day. The man sat without stirring on the riverbank opposite the warehouse.

However, the man had no choice but to eat that day, the next day, and again the following day. He was irked with the petty trade of selling baby clams, yet from that day forward there was nothing else this stranger could do. When a horse cart passed on the village path, he fancied becoming a carter, but then he had to realize that he had no money to buy a single horse. In a word, what he could do was simply use his body and, by using his body as cleverly as possible, earn, even if just a cent. Seated in the midst of reeds on the riverbank, he watched the hemp bag boat which came at least once a day, watched the young villagers crossing the plank, and thought that he also would carry those bags on his shoulders.

Carrying hemp bags across the plank, unlike what he had imagined from just watching it, proved fairly difficult. Moreover, despite the fact that it was fairly difficult, after a day's work he could not even buy a cup of sake in the evening. He had to go to the boss of the warehouse and ask if there was any other job besides. The boss said to the man: Be here first thing in the morning tomorrow.

Before sunrise, the man stood by the warehouse as the boss had told him. There was no one else. The sun's rays shone from the direction opposite the sea, and the man thought that now perhaps the boss would get up and come. The wooden door of the warehouse slowly opened, and the boss called him from inside. Line these up on the riverbank, the boss said. The man was to line up the damp, tattered hemp bags on the riverbank. As told, the man proceeded to unbind the hemp bags and line them up on the riverbank as though handling fragile things. The sun's rays already descended straight down overhead, but he still lined tattered hemp bags on the riverbank. The bags stretched way down the riverbank. The man had already lined them up to a considerable distance from the warehouse. He asked the boss in a loud voice if he was to go farther away. That's enough. Put them back where they were after you eat. Start from this end, not from your end, said the boss.

Come back first thing in the morning again tomorrow, said the boss to the man. But only if the weather is fine, he added. I'll pay you when you finish the job, said the boss and went into the warehouse. Do I finish tomorrow? asked the man in a loud voice. That'll be decided after tomorrow's job is over, said the boss from inside the warehouse.

The sun was not yet up. It'll be fine again today, thought the man as he stood in front of the warehouse. When the sun's rays started to shine from the east, the boss came out of the warehouse. Wash these in the river and line them up on the riverbank, like you did yesterday, ordered the boss. The weather is fine, said the boss to himself, looking up at the sky. Bundle them up when they are dry, like you did yesterday, said the boss. Money, and more jobs, after that is all finished, said the boss.

Waist-deep in the water, the man washed the hemp bags. He lined up the clean bags on the riverbank. When they were dry, he gathered them and bundled them up. He worked till it grew so dark that he could see neither the river water, the hemp bags, nor the boss's face. The boss gave him an unexpectedly big sum and asked him: Will you come again tomorrow? He said he would come. What do you do with them? the man asked after receiving the pay. *Susa*,* said the boss. The man did not know what he meant.

The man, having become an underling of the *susa* seller, saved money without so much as a drink. With that money, he meant to buy tattered hemp bags himself. He worked for two years steeped in the river water and bought tattered hemp bags with the money he had saved, but he did not, as the boss did, turn them all into *susa*. He examined the bags carefully, divided them into those to be turned into *susa* and those that could be sold as they were, thus creating two trades.

When the man appeared, he had some time before reaching age thirty; but as nearly three years passed since he had come to this village, he no longer had so much time. The man wanted a woman, but since his arrival here he did without one as he did without sake. If a villager asked where he came from, he repeated like a fool that he came from Yamato, but he had in fact forgotten his province. At least he remembered the mountains which he had seen as a kid. That's right, I was born in a mountain province, thought he sometimes, steeped in the

---

*Hemp or straw used to reinforce mud walls.

water. But beyond that the man recalled not a thing. Whether this man had any parents or brothers, nobody knows. What he cannot remember himself we cannot know.

The man procured an old warehouse on the river and lived in it alone. A boat loaded with hemp bags arrived, the man laid in a stock of old bags, washed them in the river and dried them as he had learned to do while working for the *susa* seller, and made a livelihood by selling them. He had become the *susa* seller's subcontractor. The man did not mingle with villagers. Were there any women in the village? he wondered. Pocketing all the money he had saved up by not drinking, he walked about the village in the evening to see. Inside houses, women all sat in front of stoves. There are no girls in this village, thought the man. He thought maybe he would have to save more money and go *next door.* He meant by *next door* the port town next to this village. He had heard a young villager at the *susa* seller's place say, crossing the plank, that there were women next door. He needed more money in order to go next door, the man thought, but he could not wait till he saved that money. Maybe this much money would do, he said to himself on second thought, and the following evening he went next door.

Next door was a place called the Warship Town. Probably, a big ship had once come to that port. In that town there were several inns, and in the inns there were prostitutes. The man entered the first inn he chanced to see and said, A woman, with this money. A woman of about forty appeared. That's not enough for staying overnight, but, anyway— she said, showing him in. Where are you from? You don't look familiar, asked the woman of about forty. From next door, said the man. Next door, you mean Denpō Village. That village, said the man. Next time you come, be a little more generous, okay? said the woman, and took him upstairs where mattresses were ready.

There was a factory upstream, as the man some time afterward heard young men crossing the plank say merrily. What does the factory make? the man asked. It's a workhouse called Hemp-Dressing Factory, said the young men. It looks like they make thread and cloth out of hemp, said the young men. The man wondered why there were women there. Are the women there expensive? he asked. Expensive? Ask them, said the young men, laughing loudly. The man had no money to go to Warship Town again.

The man no longer had time to sit amidst the reeds on the riverbank and watch the water. For quite a long time, he had not seen the sun rise and sink. The span of time between rising in the morning and dusk became short beyond judgment. No longer did he have the time to walk between rows of village houses. Nor did he even have time to turn back at a villager's greeting. Rather than become a skilled trader, or try to see if he could become one, he worked so that he would have no time left for himself. For, if he had time left, the time might wrench him down and strangle him by the neck. The river water, the rows of houses scattered on the border of the river, the thicket of reeds, and the sky were all he saw. If any spare time was left in a day, the man would have to see other things. If even a little spare time was left, the man would have to hear sounds. If he made spare time, within that time, he might be killed by the contents of that time. The man's fear was of this sort.

Is the workhouse called Hemp-Dressing Factory far from here? the man asked a young man crossing the plank laid between the boat and the warehouse. Far or not, go and see for yourself, the young men laughed aloud, and the plank shook precariously. Those men had been the same *susa* seller's employees since the man had come to this village. The factory's far, but if you mean women, they are probably near, said the men and laughed all together. How many women are there, about? asked the man. The young men said, full of cheer, How many? Well, let's see. Listen, how many, he asks. The man wondered what women did in the factory, but he didn't know the answer. What do women do in the factory? asked the man. What do women do in the factory, he says, said the young men, and one of them answered him, Spinning or weaving.

The man clasped all the money he had, and went to the factory upstream. He watched the factory on the riverside from a distance. About how many women are in there? Under the mercury-like sky the man stood. What do they look like, the women inside? How much money do the women inside there charge? Where are they from, the women inside there? The man remained staring at the factory from across the river. The man stood all alone on that land.

On the riverside it was generally windy, but the man had nothing with which to shelter himself from the wind. Besides reeds, plants like shepherd's purses grew on the riverside. In the midst of the plants, the man was alone as if he had been left behind, or dropped there in delivery. How he had lived till then, and how he would live henceforward, did not enter his mind. He waited for women to appear from the factory across the river. That was his reality then. Work on the riverside, eat meals on the riverside, and wait for women on the riverside. Perhaps these were the entirety of his experiences. The weather was fine, but he was trembling, overwhelmed by cold and the spaciousness of the land.

A few women appeared from the side of the factory. Then, several dozen more came out. Probably it was lunchtime. The women's kimonos as a whole looked like black things. Those black things went into the river and washed their hands. Those black things washed their legs, tucking up the bottom of their kimonos. Tomorrow he would go to the other side of the river, the man thought, and as he thought, the following day he waited for the women to appear near the factory and saw them from close by. The women he saw close up were much younger than the women of Warship Town. The majority were girls, and some looked like children rather than girls. When they came out of the factory, the women sat for a while wherever each of them liked on the riverside. Their faces were all dark. Whether they were dark complexioned or unwell, we do not know. If you went near them, you would not know either.

There was no place for the man to hide from the women. Despite the fact that he came to see the women, he wished to hide from them. There were, even roughly estimating, forty or fifty of them at least. He even thought of running across the river beach, but since it would be too conspicuous, he just stood on the riverside somewhat away from the women.

Suddenly, behind the man, there was a rustling sound of plants, and a girl appeared, short and dark. If you have business with the factory, the entrance is that way, the girl said to the man. I have no business, the man said. Then you shouldn't stay in such a place, the woman said, and added, still looking away, because the boss is tough. I've got money, said the man. What money? said the woman. I mean, with this money, said the man. This is no Warship Town, said the girl. If not the money, then with what, the man asked, also still looking away. The woman went running off. The women on the riverside also had gone back into the factory. The man went into the river in his clothes and swam frantically as if in a dream.

Since the women forming a black crowd called her from afar Otane-sa-a-n, the man found that the girl, short and dark-faced, was Otane-san. What in the world do you come here for, every day? said Otane-san after four or five days. I deal with *susa* in Denpō Village, said the man, and the woman said, This is Denpō Village too. Anyway, if money doesn't do any good, there is no other way but to ask you to come and be my wife, the man finally for the first time spoke what he had thought and thought for the past four or five days. Wife, you say, well, it may be better than going to Warship Town, said Otane-san.

Ah, this person was also going to Warship Town, thought the man. I was just thinking of going to Warship Town, said the woman. Over here, we all go to Warship Town one by one, said the woman. Because if we get sick anyway, we might as well go to Warship Town, said the woman. At Warship Town we can eat something nice once in a while on the customers' treat, said the woman in a tone as if she knew. I'm sixteen, but old enough now to go to Warship Town, said the woman, appraising herself. The man thought that he could not let her eat the kind of nice food that customers occasionally treated women to at Warship Town. I have kept away from sake, thought the man. If you let me eat something nice once in a while, I'll be your wife, said the woman on the riverside in the dusk where they could no longer clearly see each other's face. Ah, I'll let you eat something nice once in a while, said the man. I will have to keep away from sake again, thought the man. But, keeping away from sake, he thought, would be better than being here alone. The woman followed him.

II

Otane-san called her husband Tsune-yan. The man's name was Tsunekichi. Tsune-yan, how many times did you go to Warship Town? asked Otane-san about half a year after they married. Just once, said the man, and the woman answered, it's fine if it's just once. The woman realized she was going to have a baby and was worried about the disease of Warship Town which the women at the hemp factory had talked about. They said that Warship Town women bore rotten babies. Moreover, they said, if your man has bought a woman who gave birth to a rotten baby, your baby is also a rotten baby. Even if just once, a rotten baby by some chance may be born, thought Otane-san. If it is a rotten baby, I will just throw it in the river before dawn, she thought.

Otane-san, now living in Tsune-yan's warehouse, still had a dark face. Villagers also said that Tsune-yan the *susa* seller had brought home a dark-skinned wife. This dark-skinned Otane-san still wore a dark blue kimono. So she looked all the darker as a whole. No matter how regularly she steeps in the water every day, what's dark does not turn white, villagers gossiped. It doesn't turn white even if you soak it in *susa* bleach, they said. Before shredding hemp bags into *susa*, they now washed them with powdered bleach.

Since she had moved into the warehouse, Otane-san did the same thing every day. First she worked in the river with Tsune-yan. While he dried the hemp bags, she cooked rice. As for rice, she could eat as much as she wanted unlike at the factory, but there was nothing good to eat. However, it is not that Otane-san complained. Only, though she did not complain, she never did anything beyond what Tsune-yan told her to do. If he said nothing, she sat all day long inside the warehouse. She did, on rare occasion, mend a tear in Tsune-yan's thick cotton jacket, but she neither liked sewing nor were there enough clothes to occupy her with sewing. Otane-san did not mingle with villagers. Tsune-yan's wife is queer in the head, villagers said. That was because when, from her own carelessness with fire, half the warehouse burned, villagers who ran there saw her sit without moving. But since Tsune-yan was with her only when he worked with her, when he ate or when he lay with her, he never chided her.

I'm having a child, said Otane-san at last to her husband. If it's a rotten child, I will throw it in the river, said Otane-san. But, the baby she gave birth to was not rotten, and it was a boy. She gave birth to a boy again next year, and both were dark-skinned.

Villagers were not peasants. They were either small dealers, pack-horse men, or boatmen; gamblers or hemp factory workers. By that time another big hemp factory had been built, and many men were hired. Since they were not peasants, the villagers each lived as they liked. Most people had come to this village from other provinces because they could no longer farm, or had no land, or for some such reason. Probably, Otane-san's parents were no exceptions. By the time she was spinning at the hemp factory, they had already gone. When her father died, her mother went somewhere with a young boatman. So, Otane-san never talked about her parents, not even to Tsune-yan. Nobody ever heard her talk about herself of her own accord. But to a village grandmother who helped her deliver the babies she said, My ma had a baby too.

If Otane-san was of few words, neither did her husband speak without a purpose. That was perhaps either because they had become used to it, having had nobody to talk to for a long time, or because they had worked so long that they had no time to talk to other people. Only, Tsune-yan from time to time talked to himself; or seemed to, for in fact he was talking to something. Since it was not a person, they did not see it; and besides if they saw it, it just did not occur to them that a grown-up man was talking with the river water or riverside plants. While drying hemp bags, Tsune-yan usually chatted about this or that with the riverside plants. Otane-san, too, perhaps was talking to something. However, in her case it did not become the usual mumbling. Her occasional mumbling suddenly burst out and then ceased. It was always a curse at somebody. Since at times it was very loud, even if you, not Tsune-yan, heard it, you would have been shocked.

Tsune-yan, now with two children, had to work harder than before, but he was not as eager about work as when he had first come to this village. Just occasionally, when there were many hemp bags for making *susa* that could be sold as they were, some extra money came in. Tsune-yan, of course, still worked all day and had no spare time for anything else. No, the truth was that he had some time, but he simply had come to fear it less than in the past. During those spare moments, he watched the riverside plants. Not just watched them; at times he plucked them while talking to them. None of those plants were edible. Tsune-yan plucked grass with pointed leaves and hard stalks.

Otane, go buy a big pot, said Tsune-yan one time. Tsune-yan boiled pointy-leaved, hard-stalked grass in a pot big enough for both arms to hold. The smell of the boiling grass made Otane-san sick, but Tsune-yan boiled the grass, keeping a constant watch. When the juice dries out, white powder remains, said Tsune-yan. However, what remained after drying the juice was only the dregs of the grass. Such a good-for-nothing thing to do, Otane-san used to comment suddenly and out loud even a long time afterward; probably she was talking about this incident.

The two dark-skinned boys, as soon as they started to walk, began to play with village children. More people had settled in the village, and they were the families of hemp factory workers. Their little children sang an odd song whenever they saw Tsune-yan.

Mark, mark, pockmark
Mark to pull, it got pulled
When pulled, it tore.

Tsune-yan, that means, had pockmarks all over his face, and his brown skin looked thick because of the ins and outs. Village children perhaps had heard their parents say that Tsune-yan was pockmarked. Tsune-yan's two boys, too, copying village children as they sang, sang with them lispingly, facing him. Otane-san, after these two boys, had four girls one after another. The oldest boy, Jū-yan the idiot, as he was called, was somewhat weak in the brain. The secondborn was not, safe to say, particularly different. As for the girls, in the order of their ages, they were crazy, normal, dumb; the youngest was a diverting child, but was harelipped, and in later years one eye had glaucoma, and it became contorted after an operation.

All you do is make babies, mumbled Otane-san as if it was somebody else's business; and though the mumbling was clearly heard by other people, it was not heard by herself. All you do is make me have babies, Otane-san also said aloud to herself. It was all right while the mumbling was for herself when nobody was around, but gradually, it came to pass that even in the presence of other people, it burst out. At the village greengrocer's, Otane-san bought leafy vegetables and radishes, paid the money, and on turning back, she sometimes mumbled in a voice heard by others: All you do is make me have babies. However, Otane-san never even once grumbled face to face to either Tsune-yan or the children. She never said, Tsune-yan, please buy me a kimono or something for the New Year. Tsune-yan, too, never scolded Otane-san. By that time, Tsune-yan had a cup or two of sake at night and slept while it had its effect.

Otane-san did not pay too much attention to the children born one after another. Some of them always had a cold, a runny nose, or scabs all over the head. As the children grew bigger one by one, bigger children took care of smaller ones, and especially the four little girls looked as if they already were mothers to one another. The six children dished out and ate, as they pleased, the rice Otane-san cooked in the big flat iron pot. The six little children, just as they liked, always ladled out and ate the soup from the sooty black pot. Their mother, mute, ate rice and sipped soup by herself.

III

Summer or winter, Otane-san always sat by the side of the brazier. Fellow *susa* traders from time to time showed up, but Otane-san remained seated by the brazier, and only said as if of a stranger: Tsune-yan? If you are looking for Tsune-yan, he is at Warship Town. The couple by then had finally moved out of the warehouse at the edge of the river and lived in a little flat house in the village surrounded by streams. Jū-yan the idiot had already turned sixteen, and the second-born, Tora-yan, fifteen.

Otane-san, not long past thirty, wore as ever a dark blue cotton kimono and a plain hairdo bound in a bun in back. As a whole she looked over forty; a closer look revealed the absence of color on her face and she appeared around fifty. The youngest girl followed her around looking somewhat like somebody else's child. The dumb girl sat crouched on the dirt floor, usually looking outside.

Where's pa gone? Jū-yan always asked his mother when he came home in the evening. Each time, Otane-san answered her son saying, Tsune-yan? Tsune-yan's in Warship Town I guess. Are you going too? Otane-san said to her son.

Where has pa gone was a question Tora-yan never asked his mother. For he knew where his father was, and besides, since he had his mind on gambling which he had just learned, he was too busy finding valuables in the house to worry about his father.

The two sons helped their father, but as for the business, nothing new happened: he now stocked new hemp bags as well and barely established himself as a proper hemp bag dealer, but this was all that could be said. If you boys don't work, we may have to hire a monk or somebody, Tsune-yan said, but the two sons had learned to gamble, and learned how to filch money from the business.

Ma, we'll be back, the two sons would say and go out in the evening. The two sons, behind the mother sitting by the brazier, opened the wicker suitcase and took out things like their mother's old clothes or even men's sashes. When there was nothing in the house they could take, the two sons, when the evening came, dilly-dallied on the dirt floor.

Jū-yan, aren't you going anywhere tonight? the mother would ask the son. I may want to go but I've got no money, Tora-yan would say. Then take this, the mother would say, and undoing the threads that roughly attached the collar cover, would throw them the money sewed into the collar. You got no more, ma? Jū-yan still would ask his mother. All they do is go to Warship Town, Otane-san mumbled to herself alone. But the two sons, not bothering to hear, had run outside and were gone. Otane-san's mumbling perhaps was spat out toward Tsune-yan, though.

The two sons one day came home bare to their loincloths and covering themselves under a woman's jacket. Jū-yan was barefooted, and Tora-yan held black-lacquered female clogs. Saying, We got netted, the two quickly pulled the quilts over them and pretended to be asleep. That day, it seems, they were raided in the den and escaped by the back entrance and over the roofs to just barely make it home.

The two sons, when they made money gambling, went directly from there to Warship Town. The one who would suggest going to Warship Town was always Tora-yan. They went there for the first time when Jū-yan the idiot was sixteen and Tora-yan was fifteen. When they made lots of money gambling, they came home together toward morning. The father was usually home by the middle of the night, but occasionally he also came home in the morning. Once in a while I have to hold my wife, too; it's hard work, Tsune-yan said sometimes and laughed with fellow *susa* men. That wife, by the brazier in the dark corner, from time to time mumbled to herself, but nobody paid attention.

Ma, there's no rice, sometimes the second oldest girl would say in the kitchen. Since she had already turned twelve, she did kitchen work in her mother's place. Ma, no more charcoal, the girl would say to the mother. And the mother, by the brazier, would go on mumbling to herself as if angry: the fox in the back of the house is not dead yet. Big brother, no more rice, the girl would say to Tora-yan. She knew she could count on nothing if she had said that to Jū-yan, the oldest brother.

The oldest and mad girl, her face all white with powder which nobody knew where she got, usually stayed crouched on the dirt floor. This lunatic neither said anything weird nor went wild, so neither the parents nor the brothers and sisters thought of her as mad, and besides, this lunatic took good care of the younger sisters. She wiped the dumb child's tears and helped her blow her nose. She also dried the mattress which the dumb child had wet at night. This lunatic, unlike her mother, did not talk to herself suddenly in a loud voice. This person rarely spoke. But this person was the only princess in the house. The princess could not stand the wet mattress, and so, having no other resort, dried it.

In the meantime, Tsune-yan, the father, began to stay away from home more often. The father's trade now was carried on mostly by Tora-yan, the second boy, and Jū-yan the idiot helped his younger brother. Jū-yan, fellow tradesmen thought, was Tora-yan's errand boy. In other words, Tora-yan was not only skillful at gambling and buying women but a sharp trader. Since Jū-yan had no talent at all, he merely moved according to the orders given by his father or younger brother. If the younger brother said, How about going to Warship Town? he just followed him saying, Is it all right if I go too? By the side of the brazier, Otane-san mumbled to herself, All they do is go to Warship Town, everybody; but nobody heard her. Otane-san, looking dark, coughed little coughs, but nobody noticed. When the dumb child played near her, Otane-san lifted her up on her lap and picked lice from her hair. The child next to her had already turned five or six, but she often had convulsions. Jū-yan, this child cries so much, the mother said looking at her, but made no move to stand up.

IV

From this house, Tora-yan went out when he was seventeen. Tsune-yan the father also stayed away from home more and more; he rented a room for a woman in a place called Shikanjima near Warship Town, and now he stayed there most nights. This woman had been a prostitute at Warship Town. Tsune-yan's children called her Shikanjima aunt.

This person, it was said, was the daughter of a fisherman in Awa, Shikoku. Since she could not read, when, while still in Warship Town, a certain man wrote on a piece of paper I will kill you and handed it to her, she thought it a love letter, they say, and kept it in her purse for a long time. In passing, this aunt was only twenty-seven or -eight when Tsune-yan began to keep her, and, unlike Otane-san, was very light-skinned. Her fluffy white skin was opaque: in short, she was an albino.

Tora-yan had left home, and the trade barely went on in the hands of Tsune-yan, who was home by day at least, and Jū-yan the idiot. Jū-yan was already twenty-one or -two, but, for some reason or other, he was not taken by the army and still was a dependent at home. They had not heard from Tora-yan for nearly three years. Somehow, nobody worried much about him.

The normal girl, third from Tora-yan, was called Kine. After finishing elementary school, of her own accord this girl became a junior nurse to a doctor somewhere, and then before long passed a test, became a nurse, by some chance or other volunteered as an army nurse, and stayed at the Chinese front for some time. After that, she worked at a fairly large hospital somewhere, and as a result of becoming intimate with the head doctor's son she had a child, but they didn't let her marry him. A sum, whether meant for solatium or for expenses in bringing up the child, was given her, and she took the boy baby; she felt guilty about going home, or rather, she was concerned about that sum of money, so she rented a second-floor room and worked as a commuting nurse at a nearby doctor's place. After a while, she got together with a younger man with no job who also rented a room in the same house.

That is Okine-han's child from before, said neighbors when the man babysat for him. That child, there's something genteel about him somehow, but that's because he's of a good breed, neighbors gossiped about him. If you want to say child from before, child from before so much, I'll just marry him, and you won't have any complaint, thought Kine, and married this younger man. When big brother comes home, things will be fine, Kine thought, having confidence in Tora-yan.

The lunatic just a year older than Kine was called Kiku. This person also was already eighteen or nineteen. When she had just turned seventeen, this Okiku-san dropped a stillborn baby in the toilet. Nobody knew whose child it was. The baby uttered no birth cry, but Okiku-san cried aloud and came tumbling to her mother's side, bloody. All you people do is just make babies, the mother mumbled as usual, and brushed the daughter aside.

Really, where is Tora gone, Jū-yan thought of his younger brother whenever something happened. Kine had a child again with the new man, and on that pretext came to ask Jū-yan for money. Father is at Shikanjima aunt's place, Tora-yan's gone, where we don't know, what am I do to? Jū-yan was indeed at a loss. But, this person did not complain to his mother. From childhood, he had thought that the mother was always to be seated by the brazier. Should he say, Ma, no more money, the mother, he knew, would no longer unsew the collar of her kimono and take out money for him.

On one of those days, one of Tsune-yan's fellow *susa* dealers came.

Jū-yan, if you need money for stock, I'll lend you as much as you want, said the man. How much will you lend me? Jū-yan eagerly responded as if finding a ready boat when wanting to cross a river. As much as you want, if you just let me have a family seal, that man said. Jū-yan, you are the firstborn son of this house, you can certainly do what you want with the family seal or the like, the man said.

Jū-yan looked for the seal while his father was out, and handed it to the man. The house was mortgaged by the hand of this man, and Jū-yan, not realizing that, received from him enough pocket money for immediate expenses. Ma, I'll be back, said Jū-yan, and that day went to Warship Town for the first time in a long time.

When he returned from Warship Town, Kine was there to ask for money, having somehow got wind of it. She said that the baby this time had legs curved inward, though it was already half a year since its birth. She said that the baby would become a cripple if left like that, but that it cost money to take it to the hospital. What are you going to do? Kine stood on the dirt floor, looking as much as to say it was unjust of him not to give her money. This person did not go up to the matted floor inside where the mother sat. Of course she never entered the kitchen behind it. For this person, the house where she had been born was not hygienic. Perhaps because she had studied to become a nurse, this person respected hygiene.

Is it okay if the baby becomes a cripple? Kine spoke, still standing. Are you saying that you have money to go to Warship Town but not a cent to spend on another's child? She kept on speaking. Pa is at Warship Town and ma is crazy, they said, and so I couldn't marry into the hospital, Kine told her brother Jū-yan. Say what you will, what's not here is not here, Jū-yan said at a loss.

If I had married the young doctor of that hospital, you'd have been the one to come for money; if an idiot like you had come, I couldn't have seen you, if I had been the young madam, though. Kine kept on talking, still standing. I didn't marry into a family, I got a husband to come to me, so it's fine if I'm here to ask for money, isn't it, said Kine. That's true, thought Jū-yan. Really, where is Tora gone, he thought; he got angry at his being away whenever there was a problem.

At that point, Kiku came out from inside, and in an instant dashed across the dirt floor. Kiku the lunatic went to the riverside to have fun in the evenings. Kiku the lunatic had become a free entertainer of young villagers. The dumb child next to Kine also had turned fourteen or fifteen. This dumb child was called Kinu. Not that she heard her name called, but anyway people called her Kinu.

You cooking yet, Kinu? Kine, for the first time like an elder sister, spoke to the dumb sister crouched on the dirt floor. Kiyo does the cooking, Jū-yan answered, relieved at the change of topic. The smallest, Kiyo, was already twelve or thirteen.

Ma, I've come about money, Kine at last talked in a loud voice to the mother inside. O, yeah? Last night the fox in the back of the house cried so much I couldn't sleep, the mother mumbled. Tsune-yan's still at the river, go ask him to buy you clothes or something, the mother said to the daughter. Ma, money, money's what we need, Jū-yan said to the mother. All they do is make babies, Otane-san mumbled to herself, unheard by the daughter and the son on the dirt floor.

V

Except for crossing the bridge to Warship Town, Tora-yan had never gone out of Denpō Village. On the way back from Warship Town once, he thought suddenly, maybe today I'll go the other way. He had heard that, going straight in the other direction from Warship Town, one would come to Umeda, and that there was a station there. Past a place called Fukushima, through a place called Noda, Tora-yan successfully reached the station. There he tried to buy a ticket for Shinbashi, but in his pocket there was only enough money for a ticket to Nagoya. This and what followed were Tora-yan's own boasts after he returned to Denpō Village, so one cannot necessarily trust everything that's said.

Anyway, Tora-yan finally reached Nagoya, with not a cent in his pocket. Oh, no, what have I come to? he thought, but what was done was done; this man, Tora-yan, was a person who had no regrets of this sort. Tora-yan walked on a dark street. He walked without destination in the street of a residential area, hoping to see a garbage bin outside the servants' entrance of a house. Few people passed in the residential area. Tora-yan was smart about such things. He walked around looking for leftovers from supper thrown in the garbage. If my stomach became full, I could sleep just anywhere, he thought. However, leftovers were not as easy to find as he had expected. Every one of those well-fixed guys is stingy, he thought, while scavenging in garbage bins; then a maid came out to dump garbage from the servants' door of a house. This elderly maid saw Tora-yan, led him inside the house, and gave him rice balls, he says. Tora-yan was already seventeen, but since he was short, he probably looked like a mere child.

That house belonged to Lieutenant so-and-so, and he was a high-ranking person in the army, and I was given a grand reception as if I were Toyotomi Hideyoshi. Tora-yan, after returning home, boasted about what had happened in the house; but this person was in any case very fond of boasting.

When he arrived in Tokyo, according to him, he jumped into a pickle shop he passed and became a servant boy there. This person was somehow or other fond of pickling things and preparing food, and maybe that's why this came to pass. Hired as a servant boy, after two or three days Tora-yan already said to the master, Make dried strips from cheap radishes, they'll sell. Tora-yan had seen Denpō Village people cut whole radishes into thin strips, salt them, keep them under seaweed for flavor, and thus prepare overnight pickles. He did not see this kind of pickle in Tokyo. As expected, the overnight radish pickles sold well, and Tora-yan was treated well at the pickle shop. This was another of his boasts, so one would be on the safe side if one discounted it by half.

Starting with this pickle shop, it seems that Tora-yan moved from shop to shop more than ten times in three years. If you heard every detail of his story, you would be disgusted by so much boasting. Tora-yan never talked to others about failures and struggles.

After he married, Tora-yan was later taken by the army and sent to China, but his accounts of the front were also all stories about how he earned money in gambling, how he stuffed his belly by corrupting spineless people with the money earned, and so on. They consisted only of such absurd stories as how annoyed he was when he put the money he won in his waistband, because then his belly bulged so much. If you think I'm telling a lie, look at this, he said, and showed a picture of himself proudly astride a water buffalo. Since he did have a protruding belly, sitting on the back of a water buffalo wearing just a shirt, Tora-yan boasted all the more.

But this Tora-yan came back to Denpō Village. After an uninterrupted series of boasts on his return, he noticed for the first time that Tsune-yan, his father, wasn't home. Father is at Shikanjima aunt's place, said Jū-yan. Brother, rice and coal are all borrowed, said the youngest girl. Brother, is it right if the baby becomes a cripple? Kine said.

I see, then let's mortgage this house, borrow money, and— said Tora-yan to Jū-yan. He had learned about mortgages in Tokyo. Jū-yan, bring me the family seal, I'll find a place somewhere where they'll lend me money, said Tora-yan. That seal isn't here any more, said Jū-yan. I handed it to the old man Yamatomi and got some pocket money, said Jū-yan. Tora, you're to blame for not staying home, Jū-yan made an excuse, anticipating Tora-yan's fury.

Brother, buy me a kimono, said Kiku the lunatic. Ma told me to ask you to buy me one, said Kiku. This Kiku was beginning to swell in the lower stomach. Tora-yan had noticed it long before. What's ma doing, what on earth is ma doing, Tora-yan thought. That's right, it's no good saying anything to ma, thought Tora-yan. All ma does is grumble, he thought. Kinu the dumb was talking about something with the mother using signs.

Say, Tora-yan, don't you have money for going to Warship Town? Jū-yan asked his younger brother, already forgetting the future. That's right, I'll go to Warship Town, thought Tora-yan. Kiku will be the security, he thought. Jū-yan, not you, I'm going, said Tora-yan. Kiku, let's go to a nice place, said Tora-yan, and stood up taking the young sister's hand.

At Warship Town, Tora-yan said, This is my young sister, let her stay here for one day, and please let me have an advance for the day. Advance for the day, I have never heard of such a queer thing, the woman there said surprised, but, overwhelmed by Tora-yan's air, she handed him some money. On the same trip, Tora-yan went to a gambling den.

VI

Somehow it's been uncomfortable since Tora came back, said Tsune-yan, drinking alone, to the Shikanjima aunt. Tora, unlike Jū, is sharp-eyed, so I can't pocket anything, he said.

The Shikan-jima aunt's real name was Oharu-san, but instead of calling her Oharu, Tsune-yan also called this person aunt. The aunt was now already in her mid-forties; she grew plump as the years went by, and her whole body looked like a fluffy white lump. The roots of her arms, for example, were fluffy white lumps swaying to and fro. This person could barely cook, so Tsune-yan always had bean curd to eat with his sake. Also, since this person had not been a geisha, she could not sing. Therefore, Tsune-yan drank in silence facing this aunt. This aunt usually had patches on her temples and the nape of her neck, and said: Tsune-yan, the ceiling swirls around these days. At this rate, when I cannot see the ceiling any more, I guess I die, she said. Even if I die, I won't go back to Awa, for my parents aren't there any more, nobody's there any more, said Oharu-san.

None of Tsune-yan's children came to visit this Shikanjima aunt. Tsune-yan, once a day, wearing his rough cotton jacket, went to his Denpō Village house to tend his trade. Usually in the evening he came back there, drank sake in the aunt's company, muttered to himself about nothing particular, and fell asleep just like that. When Tsune-yan slept, Oharu-san poured in a teacup the cold sake Tsune-yan had left undrunk and drank it in one breath as though drinking water. This person did not dislike sake, but perhaps because of the habit she had brought from Warship Town, it never occurred to her that she might heat her own drink. After drinking the cold sake that was left, she had her meal with Tsune-yan's leftover rice and side dishes. Oharu-san had done this for nearly twenty years, but never even once had she thought it strange.

Tsune-yan, Tsune-yan, Tora-yan has come, said Oharu-san shaking Tsune-yan, but Tsune-yan did not wake up. Aunt, get out of here, said Tora-yan, and pulled his father up. Pa, ma's dead, Tora-yan shouted. Ma's dead, what are we going to do? Tora-yan shouted. What are we going to do, you say, but if she's dead there's no help, is there, Tsune-yan mumbled.

Otane-san, seated by the side of the brazier, had spat blood into it; thrusting her head into the ashes, she was dead. Jū-yan removed the collar cover of the kimono she wore, but found no hidden money. Otane-san died, but nobody said anything about her. Tsune-yan went back to the Shikanjima aunt's, to come home just once a day.

Sometime after the mother died, Jū-yan married a Warship Town woman who had once been a hemp weaver, and left home; he rented a room in the outskirts of the same Denpō Village and came to the house once a day just like Tsune-yan. The ones who stayed in the house were only Tora-yan, Kiku the lunatic, Kinu the dumb, and Kiyo, the youngest with the harelip.

Kiku the lunatic was expecting a baby soon. Lunatic though she was, she no longer went out to have fun; she stayed inside all day long, crouched at the side of the brazier where the mother used to sit.

Kiku, whose child, say it, Tora-yan asked, but Kiku didn't answer. Kiku, whose child, which man from where, why don't you speak, said Tora-yan angrily, but Kiku remained silent. Kiku didn't know who, so she didn't know how to answer.

Kiku was accustomed to going to the river to have fun in the evenings. Young village men came there. It wasn't that the same man always came. Some wanted to go to Warship Town but didn't have the money, and some others were too shy to go there. Kiku, being a lunatic, was convenient for a brazen man. Kiku being silent, word did not leak to other places, so even a shy man felt at ease. Kiku declined no one whatsoever's wish.

Kiku sat by the side of the brazier holding her belly so big that it threatened to drop, but Tora-yan was not at a loss about her having a fatherless child. He was at a loss in the same way as he would be if a dog or a cat had many babies until there was no more space. This Kiku, if left alone, might have many babies like a dog or a cat. In short, Kiku had nothing else to do except have babies. After dropping a baby in delivery in the toilet, she already twice spurted a lump of blood.

One day, Tora-yan saw a small man, lean and thirtyish, standing in front of his house spying inside. The following day, and the day after the following day, this man without fail stood in front of the house in the evening. Hey you, what do you want here, Tora-yan finally asked the man, grabbing him by the chest. I thought Okiku-san might belong in this house, the man mumbled in a small voice. So what, said Tora-yan, furious beyond reason, threatenting to beat him at any moment. If Okiku-san belongs here, I thought I'd ask her to come and marry me, said the man trembling. What, are you the one that made her belly swell? Tora-yan dragged him inside the house.

This man, they found out, was a horse packer already past thirty, still a bachelor. Since I've done that to Okiku-san, if she is having a child because of that, there is no way but to have her come and marry me, the man said. That's right, because you did such a thing, Okiku's belly swelled, said Tora-yan. That's right, thanks for coming, because Okiku is having your child, sure she is your wife, said Tora-yan. Then let me take her home please, the man said, and took her away almost dragging her, quickly so that he would be able to escape before Tora-yan hit him. What's this, that man, isn't he a grown-up? What an odd fellow, said Tora-yan tut-tutting.

At any rate—said Tora-yan to himself. He was worried at heart, and his worry was about the mortgaged house. He did not know whose house it was now at all, but he had to get back the family seal from the old man Yamatomi. That old man is a scamp, so I've got to make money first, thought Tora-yan in resignation, yet he could not think of how to make that money. Petty gambling could not of course instantly produce a sum big enough to restore a house.

Tsune-yan, the father, who showed up once a day, no longer was at all interested in the trade; far from it, he had no desire to make money. This person by nature did not like mad pursuit to begin with. Therefore, this person could not forcefully take money even from his son Tora-yan; when he saw Tora-yan, he could only say, how's the business, but, since Tora-yan was smart, he handed some money to his father.

Like this, Tora-yan is strangely generous toward his family. About the older brother Jū-yan's mortgage matter, he never grumbled even once afterward. At heart he always worried about how to make money for it, but since this was nothing to boast of, he didn't speak about it. I want you to recall that he was fond of boasting.

Since Jū-yan, the oldest son, married out of the house, Tora-yan, though the secondborn, was treated by others as the oldest son and heir. Jū-yan, the oldest, showed up like the father once a day at Tora-yan's house, subcontracted Tora-yan's trade, and lived on the pocket money thus earned. This person, called Jū-yan the idiot, could not handle trade on his own.

It worked out all right, because Jū-yan's wife managed housekeeping on his small earnings, miraculous as it was; and Tora-yan said, impressed: How can you live like that at your place, you have no candles so you light your own nails. Anyhow, Jū-yan's wife never threw away anything named food, whether rotten, or musty, or nibbled by rats; nor did she even show up for a next-door funeral, much less mix with neighbors. Nor did she ever appear at Tora-yan's place. Perhaps she figured that her clogs would wear out if she went out. This wife, dark, short, then, really resembled Otane-san who had died, though she did not mumble to herself.

VII

Tora-yan, as ever, liked gambling, quarreling, and whoring, but, already twenty-seven, he carried on his trade in an easygoing way. However, Tora-yan had no wife yet. Well, the fact is, though he was a rather able tradesman, since his passion for gambling was well known in the neighborhood, it was not strange that nobody came to marry him. At least, no decent girl could be expected to come.

Besides, this person did not like being told the truth. There was in the neighborhood a decent girl who had an eye on Tora-yan, but when once this girl said to him that gamblers were human rubbish, he knocked her down, and the girl disappeared from his sight.

By then Tora-yan's Denpō Village had become Denpō Town, but Tora-yan's house stayed the same as before, only older if anything. Inside the house there were, besides Tora-yan, only the dumb child, and her younger sister Kiyo. This Kiyo after a while went somewhere to serve as a maid, so Tora-yan at this point was alone with Kinu the dumb.

Tsune-yan, the father, kept away from the house all the more; he came just once in five days or so, got from Tora-yan a meager sum for food, and went back. Rarely, if Tora-yan was alone drinking unheated sake or something in the kitchen, he said, Drinking, eh? and joined him for one cup, but went right back. Half of Tsune-yan's hair had turned white, his face and the nape of the neck were reddish brown from drinking.

About that time, Tora-yan engaged in trade which consisted of buying used machines disposed of by hemp factories as waste iron for a few cents per kilogram and selling them to scrap iron dealers. By the ordinary procedure, there was no easy gain; but Tora-yan profited because he had the knack of fixing the scale when weighing discarded machines, and the skill of getting them on a low bid when there was bidding for a large quantity. Before bidding, Tora-yan bribed the man in charge. And yet when he sold the machines, he was not willing to come down even one cent; so he was reputed among his fellow tradesmen to be tight about commissions. Moreover, though they were called factory discarded machines, some could still be sold as they were, so he sold them at machine prices, and there again he wouldn't come down. If one wondered if he enlarged his trade with the money earned this way, or started a new one requiring investment, Tora-yan did not seem to do that sort of thing.

Kinu the dumb was at an age when she could have married, but this person could do no kitchen work, no needlework, nothing. So Tora-yan brought home fish from a fish store, cooked them, and drank alone. But usually he went out to drink in the evening. When his business allowed him, he went either to gamble, or to play billiards which he had just learned about that time and had a passion for. His billiards, of course, was on a gambling basis. This person didn't get excited about anything unless money was at stake. After he had married, when the first child was being born, he heard his wife groan while seated on a bench in front of the house with some men in the neighborhood, betting whether it would be a boy or girl.

Tora-yan, though reputed to be a skinflint, possessed a strangely effective narrative skill and was, moreover, generous at a place like a bar; so he was popular among women in such places. However, that was just at the beginning, and since he was sordid about money in an odd way, women did not last long, either. When he met a woman to his taste, this person, even if he had to borrow money, gave her an expensive kimono or anything she wanted; but, once she became close, and when his wallet was light, he had no scruples, for example, about borrowing money for cigarettes from her till he disgusted her. Within the past two or three years, Tora-yan made pregnant a bar woman and a woman at a cafe newly started on the other side of the bridge; he didn't even pay them half the amounts they demanded and ignored the consequences.

In one of those years, a big storm struck, which turned out to be a typhoon, as the Denpō people later learned. Denpō Village, or rather, Denpō Town, as it was already known, was near the mouth of the big river, and this river, branching off into many smaller streams, surrounded this land like an islet; so, a big storm brought flood to this land.

Suddenly with a rapping noise raindrops fell, and doors and signboards here and there hurtled into the air. Zinc roofs were ripped off like pieces of paper, and roof tiles created a snowstorm. Already by that time something dark was progressing from beyond the bridge, and nobody ever thought it could be water. Water! people shouted, but by then the dark water was up to the armpits of the Denpō people, and before they got up to the rooftops, many people had been sealed in by water up to the ceilings. Such a thing as a two-story house did not exist in the town of Denpō, not even one.

By chance, in Tora-yan's house, the water did not reach the ceiling. When it was clear that the water would not rise above his navel, Tora-yan, from the storage room by the side of the house, took out four empty gasoline drums. Put the plugs in and use them as floats for Kinu the dumb and himself was his calculation.

Some old people were put afloat in wooden tubs to take refuge on top of the roof of the hemp factory. Like Tora-yan, people waded almost swimming toward the factory, with children holding on to empty drums. Tora-yan, too, holding Kinu's hand, stumbled across the water toward the factory, but since Kinu could not walk well, in the end Tora-yan swam, holding her sideways. Tora-yan, having swum across the river, they say since age five, was a good swimmer, but this was different from swimming in the river.

More than when the water came, people suffered when it receded. Those who were sealed to the ceilings were dead when the water subsided. Even most of those who survived grew sick. Seeing that the water was going, people threw things in it. Tatami mats floated away like rotten bean curd. Unless they threw them away, people foresaw that they would be stuck with waterlogged things. In the water, food and excrement washed away together. At Tora-yan's house, too, what remained edible after the water had gone were only the very bottom layers of the pickled plums tightly packed in a big jar. Those plums, it was hard to believe otherwise, were surely picked by Otane-san, but if so, it is really strange. It is a sort of miracle.

When things settled down, water gone, Tora-yan first ran to the rice store in town. The idea was to buy water-soaked rice. It was to buy the relatively less saturated, toward the back of the piled-up bags of rice. But if he said such a thing, they would refuse to sell them; so he bought all the bags in the store for next to nothing. He also bought up water-soaked rice at rice shops in neighboring towns. This rice Tora-yan spread to dry on the street in front of the house, now clear of water. The rice that had absorbed water, when sunned and aired, naturally fermented. For a while the street in front of Tora-yan's house was filled with a smell suggestive of unrefined alcohol. Tora-yan sold this dried rice to the victims, who could not buy expensive rice, at half the price of regular rice and made a killing. Was Tora-yan greedy?

In this flood, neither his father nor brother nor sisters died.

Tora-yan, who made money due to the water, after a while brought home a wife. This person had been an entertainer-maid in the pleasure quarters or some such place. She was a year older than Tora-yan. Probably, Tora-yan had visited the pleasure quarters with money. And yet since he was not the class of patron that would redeem a geisha, he probably became intimate with a maid. This person had three children.

Tora-yan's wife was called Kotoe. Before coming to Tora-yan's place, Kotoe-san had left the nine-year-old boy with her mother, and had the five-year-old boy adopted by a neighbor, but brought with her the six-year-old girl. Tora-yan knew all this. Kotoe was then twenty-nine.

Kotoe consented to become Tora-yan's wife, partly because she was attracted to his disposition that shone only at the beginning, but the truth was that she wanted to cleanse herself of the entertainment trade. Kotoe had formerly been the daughter of a decent merchant, but when she was of age, her parents' business failed, and to save the situation, she was married off at age nineteen. Her oldest son was a child by this husband. When she left this husband, she became a geisha, saved by the childhood training in entertaining arts her parents had provided her for their own pleasure.

Because she wanted to quit being an entertainer, by and by Kotoe had a family with a man. The other two children were this man's children. Although they had children, this man did not register them. This man, having evaded the draft, was supposed to be dead and had no family register. When they had two children, he suddenly vanished with no notice, and Kotoe, in order to feed the three children, again made use of her training as an entertainer.

See, as I told you, Tora-yan's wife was from a certain place—neighbors talked about her when Kotoe went to the bathhouse. In the town of Denpō, nobody went to the bathhouse like Kotoe, neatly wearing a regular sash over a *yukata.* Nobody had a proper hairdo every day like Kotoe. Your place has become quite uncomfortable, brother, I cannot even eat here relaxed, complained Kine who came once in a while. That was because Kotoe served food individually in plates and little bowls. Kotoe moved the rice from the pot to the rice server, but this had been rare at Tora-yan's house, and was felt too formal. Tora, isn't your old woman tight, she doesn't even serve a cup of sake, Tsune-yan, the father, also said. Nor could Tora-yan drink unheated sake from a teacup while standing in the kitchen. Before, those who wanted to eat ladled rice and soup from the pots when they wanted to eat, and ate as they liked; but now they couldn't do that. Tora-yan, by now, was given his share of clothes according to the four seasons.

One year after Kotoe came, a girl was born, and scarely had he seen her face when Tora-yan was taken by the army and sent to China.

VIII

A while after Tora-yan came back from two years in China, Tsune-yan, the father, died. On his bumpy, pitted, brown face one day a small white pimple-like growth appeared. The next day, on his bumpy hands the same small pimple-like things came out. Since they were neither sore nor itchy, he did not worry about them, but those little things appeared every day somewhere, and before he knew it, they covered almost the entire body, breaking one after another and oozing water. Tsune-yan's bumpy, pitted, brown body was full of little growths, and as they broke, it was nearly impossible to tell dents from bumps.

As Tsune-yan became full of growths all over his body within a month, Tora-yan, the son, startled, took him to the hospital; but before a week passed, Tsune-yan died, his whole body wrapped in white bandages. When he died, all that was visible from outside were the two eyeballs.

When Tsune-yan died, Oharu-san, the Shikanjima aunt, went back to Awa and, it was said, she spent her days working for a fisherman, carrying fish or mending fishnets. Oharu-san, with the fluffy white skin, used to say she was dizzy around the time Tsune-yan died, so she probably died eventually, falling on the seashore, or in a fisherman's hut, somewhere in Awa. Nobody wondered whether the Shikanjima aunt lived or died, however.

While Tora-yan was away, Kotoe hired a Korean and carried on Tora-yan's trade; so when he returned from China, he could have resumed the trade right away. Moreover, during his absence, that matter about the mortgage on the house was all cleared up, and the old house was newly rebuilt. In addition, Kotoe handed Tora-yan a bankbook showing a considerable sum.

Yet Tora-yan did not put his hand to the trade; instead, every day he went out to have fun somewhere with some money in his pocket. It seemed there were women, of course, but he also appeared to be possessed by horse racing. Thus the savings disappeared in the twinkling of an eye. Kotoe and the Korean, as during Tora-yan's absence, worked all the time. In the midst of it, Kotoe had a boy baby. Ten days after she had the child, Kotoe again started out to buy scrap iron or to bid at a hemp factory. Two weeks after she had the child, Kotoe hired a rough day worker to carry the big machines. Even if she wished to quit, since Tora-yan did not stay home much, Kotoe could not just leave the trade alone.

But, Tora-yan, when home on a rare occasion, lectured Kotoe and the Korean: When you were sleeping in this kind of place with your navels facing the sky, I was under gunfire. In China, women and children like you are substitutes for straw effigies. When their heads are half-off, we stay watching on this side to see if they still walk, like cocks. It's a mess, really.

Since he was dead, Tsune-yan did not come, but Jū-yan still came as before. As usual, it was to get a share of a job that would bring money, or in the event that there was no job, to beg for money. Kine, who worked as a nurse, rarely appeared at Tora-yan's place. It was likely that, behind Kotoe's back, she got hold of Tora-yan outside the house and begged for money. Kiku the lunatic, since she lived in the same town, at times dropped in unannounced, but this person had no purpose. She already had five children, and strangely, all five were normal.

The youngest, Kiyo, who had gone out to work as a maid, married the truck driver at the place where she served, and lived in the next village. This Kiyo showed up at Tora-yan's place only when she wanted money. Now she was with child, now her husband was in an accident, now he wanted to become a mover—and each time she came to Tora-yan's house. What happened to Kinu the dumb? Before people knew it, she also had a family in the same town. The husband was a tall, stoutly built longshoreman. This Kinu, unlike the others, never came even once to Tora-yan's house after she married.

Besides his brother and sisters, Kotoe's mother had moved into the town of Denpō. This person was thickset and fat and always with a peaceful smile pulled a cart selling dumplings. Needless to say, she was taking care of Kotoe's first child, her grandchild.

There was yet another person who came to Tora-yan's house on occasion. It was Kotoe's older sister who lived alone at a certain distance. This person was older than Kotoe by almost a dozen years, so she was past her mid-forties but, bereaved of her husband three times, had no children. This person's hobby was pleasure gambling. On this point, she got on well with Tora-yan. Tora-yan, therefore, went to the sister-in-law to borrow money for gambling. Then, this sister-in-law, unlike Kotoe, handed him money anyhow, saying that it was unpropitious to criticize money for gambling. Kotoe would pay no attention, saying, I'd rather die than see you gamble.

This big sister, named Takano, from time to time became possessed by a god. Once, when she was visiting Tora-yan's house, she suddenly started to pray. She joined her hands, raised and lowered them, and held them above her head. While she repeated this, her body started to tremble, and a god entered into it. When the trembling subsided,

Takano said to Kotoe: In back, there is a white fox. This fox asks to be worshiped. His name is Chōkichi-san. Make a shrine for him beside the entrance.

Thus there was a cheerful shrine beside the entrance to the dirt floor to Tora-yan's house, and morning and night Kotoe offered candles. And again on another occasion Takano, at Kotoe's house, showed the shed skin of a snake about two meters long, saying, There was a big snake in this house a long time ago. Look, I found this skin in back, and this snake is called Yonekichi-san. Thus, on both sides of the shrine at Tora-yan's house were placed ceramic flower vases in the shapes of a fox and a snake, one marked Chōkichi and the other Yonekichi. Besides these, it enshrined the fox deity, god of traders, and also the stone idol whom Kotoe worshiped every month for the sake of her boy in ill health; so in Tora-yan's house, there were enshrined many gods.

On top of this, while Tora-yan was away in the army, Kotoe bought a fine Buddhist shrine, almost incongruous in this house, and there enshrined Otane-san and Tsune-yan. On rising in the morning, Kotoe first offered light and water to the gods and the buddhas. In the evening, too, before supper she offered light, water, and fresh cooked rice; and once in three days, she gave fresh flowers.

Once when Kotoe was offering candles to the gods in the evening, suddenly the tail of the fox-shaped ceramic vase for Chōkichi-san cracked and fell. Right at that moment, the Korean employee ran in looking pale, saying that he had just run over an unfamiliar little child with the scrap iron truck. However, the child whom he thought he had killed was safe since the wheels had hit no fatal areas; his only injury was a limp in one leg. This Kotoe interpreted as due to Chōkichi-san who had saved the child by sacrificing himself. Tora-yan joined his hands toward neither the gods nor the buddhas. This must have been one of Kotoe's dissatisfactions and anxieties.

After a while, Kotoe's mother, the dumpling seller, died. This person, it was said, was too plump to be put in a coffin.

IX

Ten or fifteen years, or maybe twenty years, later, Tora-yan's country was at war with some foreign countries. The war lasted some years, and in the end the house Tora-yan lived in was among the ones burned down. From the sky things called incendiary bombs fell. The scrap iron that Tora-yan dealt in had to be turned over to the state, leaving not even a single nail. Tora-yan's business fell off. Moreover, the houses in the neighborhood of the hemp factories were torn down. They said they were making parachutes at the hemp factories. Tora-yan's house, too, was pulled down, so the family moved to the outskirts of the town of Denpō.

Hit by incendiary bombs, the flat blew away in pieces, and Jū-yan and his wife died. Tora-yan dug out the corpses of his big brother and his wife and, putting gasoline to the two black bodies, burned them on the spot.

The children of Kiku the lunatic, all five of them, were burned to death while trying to run away from the incendiary bombs. Okiku-san, it is said, fell in the river while fleeing the fire and died.

Kine, though her house was burned, lived, and burst into Tora-yan's house with her husband and two children. This person had good luck. The long flat where Kinu the dumb had lived also burned, but she made a shack on the outskirts of the town of Denpō and stayed there. But, the husband, a drunkard longshoreman, later on became palsied and seemed to have lost the freedom of half the body. And how Kinu was eating, nobody knew, but nobody had the leisure to pay attention to such a thing. After a while, Kotoe heard a rumor that Kinu the dumb had turned blind. Dumb and blind, she probably died somewhere, unable even to curse.

During the war, even Tora-yan lived by growing sweet potatoes. At that time, his livelihood was no different from most others', and he spent his days looking for food. The war ended in defeat, but Tora-yan quickly went back alone to the town of Denpō, started on the ruined field his old business, and in a year or so this turned into an ironworks. Tora-yan went back to the town of Denpō, but Kotoe and the children didn't. Or rather, Tora-yan didn't go home to Kotoe and the children.

That was because Tora-yan started to live with a woman who ran a small drinking shop in the black market. Kotoe and the children continued to live on the outskirts of town. Kotoe reviled Tora-yan, saying, Your father's blood's in your body, you act just as he did. Otane-san grew mad and died because she was jealous about the Shikanjima aunt, but I won't die, said Kotoe; even so she took to bed a few times, sick at heart.

The year after the war ended, Kotoe, for the first time after eight years or so, and somewhat past age forty, had a boy baby. It was Tora-yan's child. No matter how debauched, one can't lay a woman while the bombs are falling, Kotoe said. After three children left for school, Kotoe usually sat on the side of the brazier. By then, the girl Kotoe had brought was no longer there; married, or something.

Kotoe, when somehow set off, complained to the three children. I toiled and toiled, and we just reached a point where we could relax a bit, and look what's happened; who made a man out of that gambler? The children did not listen to Kotoe's complaints, but once the impetus was there the complaint went on and on: Isn't that right? When he was in the army, I worked till I was all black, and when I thought I could finally save some money, he came home, and wiped it out clean. And he got hitched by that whore-like woman. What's he going to do now, really?

To the girl just turned a middle school student, the primary school boy, and a nursery school boy, Kotoe pleaded, crying. With that kind of woman clutching his balls, he dares get stingy about his own children's food money. As Kotoe cried bitterly, the children, each in a different way, vaguely felt the fear of starvation. Then, the smallest child cried aloud, and the loud cry, which the boy could not suppress at once when slapped by Kotoe, lasted endlessly. Turning small like the sound of a pipe, it irritated the mother and the older children all the more.

The oldest girl, after finishing school, after a while eloped somewhere with a schoolteacher who had a wife and a baby on the breast. At the back of the girl leaving with a single suitcase, Kotoe shouted: Your father's blood's in you, chasing a man's butt. All right, do what you want.

X

Tora-yan slept on the second floor of the small tavern in the black market and from there commuted to the ironworks in the town of Denpō. The black market turned into a shopping area, and the small tavern became a mahjong parlor, but Tora-yan still lived with the woman on the second floor of that place. The woman, though younger than Tora-yan by a dozen years, was skilled at business. Not just at business, she was good at handling a man like Tora-yan, it seems. Tora-yan was fond of boasting and hated being told the truth. This woman never told Tora-yan the truth. A migrant from somewhere in Kyushu, this woman had started her shop in the black market.

Once when Tora-yan went back to the second floor of the mahjong parlor, this woman was in bed with a young man he didn't know. Tora-yan had passed a few years beyond fifty. Tora-yan just went downstairs, but came again with a cooking knife. There is no knowing what happened upstairs, but it is said that Tora-yan had to write an apology at the police station. However, except for this woman's second-floor room of the mahjong parlor, Tora-yan had no place to stay. He went on sleeping there after that event.

Once in a long while, he had a phone call from his daughter at his ironworks, and it was invariably for sponging off him. The daughter apparently lived in an apartment somewhere with the schoolteacher, but she did not tell the father its whereabouts. In response to his daughter's call, Tora-yan met with her at a sushi shop somewhere and drank beer sitting side by side at the counter, feeling uneasy. That the young woman drinking beer beside him was his own daughter somehow or other felt odd. Therefore, he always said with emphasis, Don't tell mom; then the father held out the money and the daughter took it.

Tora-yan nearly committed suicide. Nobody knows the reason. Whether it was because of a woman, or a business loss, is not known. What method he thought he would use is not known either. That's because Tora-yan did not tell. However, after a considerable time, when he said, I nearly died but couldn't die, his daughter didn't think it was a lie. Tora-yan was fond of boasting and never talked about his failures, as everybody knew; so when Tora-yan said this, nobody who knew him believed him. They thought he was joking. Kotoe, too, only said, Stupid, how would that man die? When Tora-yan said, I nearly died but couldn't die, the daughter joked: Isn't that because you've grown useless that way and you are no longer a woman's favorite, father? Tora-yan said, How can that be, and laughed, but the daughter thought she could believe that he had nearly died.

I thought of dying, I nearly died, Tora-yan said two or three times to his daughter, but he continued to live. The village where Tora-yan had been born was now a town, but was not much different from when it was a village. Only, besides the hemp factories, many and various factories had been built because there was the river. Perhaps for that reason, the river somehow reeked of chemicals. When one entered town, there was a feeling that a sticky sweet smell of gas hovered over the ground. On the banks of the little streams, junk shops stood side by side. They were subcontractors of scrap iron middlemen like Tora-yan was long ago. Old gasoline drums and rusted old machines were dumped there in disorder. On the bank of the big river, big factories stood closely together, and from chimneys dirty yellow smoke and purple smoke was spat out. The plants on the riverbank had been buried by concrete.

Tora-yan did not die of his own accord, but of illness. Each time he drank sake, he pressed his side and said, There is a lump here; it was a lump of bad flesh that grew bigger and bigger. In order to take it out, he had an incision made in his belly, straight and as long as a foot, but the lump of flesh, already intertwined by strands of blood, could not be removed, and with the lump of flesh still nesting there, Tora-yan died. From about three days before he died, Tora-yan cried all the time while he had consciousness. I don't want to die, I don't want to die yet, he kept on crying. Tora-yan lived till he became skin and bones, and when he was skin and bones, he finally died; but Kotoe, when he was about to die, went home from the hospital, saying, I don't want to see.

Kotoe is still alive ten years after Tora-yan died, and from time to time, when she is set off, still speaks ill of Tora-yan and complains, but nobody can silence her. Kotoe-san is a person of faith, so she probably won't cry when she dies, but she repeats to her sons every day: My children are all grown-up, and I don't have anything to do now, but even so it's not that I can die—besides it is scary to die.

Kotoe's sister Takano is not dead yet, either; but since this person is already eighty-something, she says, I am ready to be called any time, and at the nursing home, she eats the festive breakfast neatly dressed on every New Year's Day. On New Year's Day a few years ago, Kotoe's son, given away somewhere to be adopted a long time ago, called her on the phone. You may have forgotten, but I resent you, said the son, it is said, and the phone clicked.

十一

Congruent Figures

Takahashi Takako

I went out to get the mail. Numerous persimmon leaves were scattered on the long stone pavement leading from the entrance of the house to the gate, covering it almost completely.

The autumn wind had blown all through the night, and the voice of the forest in back of the house, as if weeping or panting hard, continued to stir the deep layers of darkness. In the evening, before the wind started, many winged insects crept inside through the cracks in the windows and crawled across the yellowed tatami mats. They kept coming in even though I killed them continuously. This morning, however, the dark, clear sky spread out silently.

There are three old persimmon trees alongside the *karatachi* hedge. The fallen leaves had been blown to the side of the building rather than to the side of the hedge. The leaves, clearly showing intricate combinations of yellow and red colors, were piled atop one another, each of their individual shapes clearly distinct. They were so vivid as to appear to have taken on color even after falling to the ground, for they did not impress me so much while they were on the trees. I stepped on them carefully. I felt they were too good to step on.

In the mailbox there were two letters for our son, one for my husband, and one for me—while I was checking the letters like this, I found that the handwriting on one addressed to Mrs. Akiko Matsuyama evoked a strange feeling in me. It was like the feeling evoked at the

---

Sōjikei (1971). Translated by Noriko Mizuta Lippit with the permission of the author.

moment you receive an envelope addressed to you written by yourself without realizing it is your own handwriting. I have experienced several times before the sense that something with which I was already too familiar was emerging from the handwriting, and gradually come to the realization that it was indeed my own handwriting. But this was not a return envelope and the handwriting was not mine. It was not necessary to turn over the envelope to find out the sender. The way the pen was used—the tip of the pen pushed hard on the paper as if it were a thumb tack, the precise shape of the letters written with unfailingly careful strokes, and then toward the end of each line, as though the tension were suddenly released, the style becoming wild and carefree—it was a handwriting with the same characteristics as my own. What does Hatsuko have to say, Hatsuko from whom I had not heard for a long time. I opened the envelope in front of the gate.

Mother, how have you been? I imagine both father and brother are fine. I have not seen them for more than four years. Not since you came to attend my wedding. I now have a child—an infant only a year and three months old. Will you scold me for not letting you know about my pregnancy or the childbirth? No, I don't think so. Even if I had let you know, you would not have come.

Have you ever taken a Noh mask in your hands and looked at it? It is very strange. When I hold the mask in both hands and gaze at it from the front, it does not have any expression. Yet shifting the mask a bit to look at it from a slightly different angle makes some emotion appear vividly on it. But it soon becomes absorbed into the face and the face returns to its original expressionless state. Upon shifting the angle again, some other emotion emerges on it, but it too coagulates into a hard, motionless expression. Each time the angle is changed slightly something emerges, but its existence is only hinted at and the mask returns to its former expressionlessness—indeed, the Noh mask is strange. Do I offend you? Will you take it as the offensiveness of a daughter who was disliked by her mother without knowing why?

It was a long time ago when I became aware of that face of yours. The whole family, all four of us, went to the beach in the summer when I was in the fifth grade. My brother and I sat in the back of the boat and you sat in front while father sat in the center, rowing. Do you remember father taking us for a boat ride? I came to realize that your large eyes were fixed on me across father's shoulders. I wondered why you gazed at me in such a way. Your face was like that of a Noh mask. As the boat swayed, your face tilted slightly and a certain vivid emotion seemed to appear on it, but your face kept its overall expressionlessness. I could not stand such a gaze and stood up suddenly. Because of it, the boat almost turned over and I fell into the ocean. Do you remember? I sat down again in the back of the boat after father and brother helped me back in. My eyes hurt because the sea water got into them, and the tears continued to flow. This time I looked at my mother through a veil of tears. Then you quietly looked aside, showing your pale profile, and from that time you continued to gaze vacantly far away, looking into the open air.

From around that time, you did not talk to me frequently. What were you angry with me about? I wondered. Since you were the same old mother to my brother, I thought for a moment that you were merely maintaining our generations-old family custom of treating boys with respect but bringing up girls with strict discipline. Sometimes you looked at me with a hard face devoid of emotion, and after that you always looked aside coldly. Since I became disliked by you without any apparent reason, I could not help but move out of the house as soon as I graduated from high school by choosing to attend a college far away from home. No, rather it was mother who strongly suggested that I leave home.

I must have written something which will offend you. This may serve as my apology for having been absent from home for so long. Next Thursday, October twenty-fourth, I will visit you after such a long absence. Since it was soon after my graduation from college that I last went home, it's been six years since I last returned. The soil which retains the familiar smell—but you are standing on it as if to prevent me from feeling nostalgic. Since my husband must make a business trip to that area, I will accompany him and will stop over, just me and my child. Please take a look at your first grandchild.

October 18
To Matsuyama Akiko From Ino Hatsuko

I folded the letter and put it in my pocket. I returned to the house, slowly stepping on the persimmon leaves on the stone pavement. The color of the fallen leaves, the delicate mixture of yellows and reds, did not impress me as vividly as it had before. It was because something still more vivid, something like many colored pictures, came sliding into my mind.

"I see. Hatsuko too had a child," I said to myself in a low voice. I wondered whether it was a girl.

Was it something special that I felt about Hatsuko? Was my feeling about her abnormal? No, I don't think so. It was an emotion which all the mothers of this world must have felt about their daughters. For most mothers that emotion might have been like small white bubbles coming up every so often from the swamp that is called life, where the corpses of their ancestors, related to them by blood, are lying about. Although mothers sometimes recognize those bubbles, they will burst on reaching the surface of the swamp, disappearing in the open air. I must have lived that emotion as if I had enlarged it through a magnifying glass.

I could imagine vividly in my mind's eye the color of the sea's surface on that summer's day when the boat was gliding smoothly along. It was a quiet sea after the tide had ebbed and before the new tide swelled. There were no high, coiling waves, but flat thin ones, widely separated, which formed a stripe design. The waves were constantly moving toward the beach, but because their movement was slow, or because the same shapes were repeated, I had the illusion that the sea was immobile with the stripe design shining on it. The dizzying profusion of lights spread as far as the eye could reach, and when I looked into the depths of the water, I found only a dark green stagnation hidden there. Although I was born in a small city away from the sea and married into a family living at the edge of the same city, it was not my limited opportunities to see the sea that kept its color that day in my memory, distinguished from the color of the ordinary sea. It is natural that Hatsuko remembers my face on that day, as she wrote in her letter.

I recalled several old Noh masks which I had seen once at the temple of one of my relatives. They were hanging on the wall of a dark room with a wooden floor, a room which was supposed to be a treasure storeroom. They seemed to be floating in midair in the dark room filled with the smell of dust and mold. I remember quite well the Shakumi mask which expresses the middle-aged woman. A white, smooth mask. While the mask looked as if it were smiling, sad, angry, afraid, or mad, it neither smiled, nor was sad, nor angry, nor afraid, nor mad. The mask itself was expressionlessness. The reason why it had to be so expressionless is that it contained overflowing emotions inside.

It is true that I was watching Hatsuko who was sitting in the back of the boat for a long time from the prow of the boat. Yet I have to go back to earlier experiences to make myself understand the feeling I had in the boat.

On a spring evening when Hatsuko was a third-grader, Masao and Hatsuko went out the gate as usual. The children called the time around six-twelve or -thirteen father's time. My husband was working for the technical division of a textile producer with a large factory; the company was located on the other side of the mountain which borders this small, provincial city. Since the train ran every thirty minutes and the bus every twenty minutes in those days, unless the transportation was delayed, he always returned home precisely at six-twelve or -thirteen. I heard the voice of my husband mixed with the voices of the children coming from the direction of the gate when I was preparing supper. I used to hear the sound of these three voices coming always at the same time in the evening as if receiving a sign of the sureness of life, if calling it a sign of happiness would be an exaggeration. With their voices as a signal, I came out from the front to the stone pavement. The three persimmon trees had spread their new green leaves alongside the *karatachi* hedge and the air around seemed to be tinted green. I thought it must be because of this that my husband's face looked dark, but even after he passed through the shade of the new leaves and came in front of me his complexion looked somewhat different from usual. I tend to notice any subtle changes of facial complexion whenever I look at anyone's face. Yet it was not so bad as to make me ask him why.

It was the habit of my husband to enjoy his supper, taking a long time. But that day his chopsticks moved particularly slowly. He seemed to be very tired. I sat facing him across the square table, and Hatsuko and Masao sat facing each other.

"Shall I give you a piece of cicada tempura?" So saying, Masao picked up a shrimp tempura from the large plate with many pieces of tempura and put it on a small plate for Hatsuko.

"I will give you a piece of worm tempura." Hatsuko picked up string beans from a large plate and put them on Masao's plate.

While eating I cast my eyes vacantly on the movements of my husband's chopsticks. The black-lacquered chopsticks approached the tempura platter, hovered uncertainly, approached the *tsukudani* fish and picking up one small fish brought it slowly to his mouth. He chewed it as if chewing a toothpick. Next the chopsticks were extended again, became uncertain, and finally chose some vinegared spinach. As I was thinking that he did not have any appetite, he put down the chopsticks together with his rice bowl, which still had unfinished rice in it. I tried to say that he did not look well, but I served him tea first. Yet it looked as if he had not finished eating; he picked up the chopsticks again and absentmindedly looked at the table filled with plates, bowls, and teacups as if looking for something. Thinking he was looking for a pickled apricot, I stretched my hand toward a small bowl containing them. At the very same moment, Hatsuko's hand reached out to them.

"Father, isn't this what you want?" Hatsuko's hand picked up the small bowl of *umeboshi,* leaving my hand dangling helplessly in midair. I saw that her hand was exactly like mine in shape and color. If her hand were enlarged, it would have become the same hand as mine.

"You don't look well today, Father," Hatsuko said. It was just what I had intended to say.

"That's not true. Look at him carefully. He looks the same as usual." I felt compelled to say it.

"Hatsuko may be right." My husband cast a weak smile at me.

I should have rejoiced in the sensitivity that a daughter of only nine had just displayed. But instead, some unexplainable feeling of minding it stayed in my mind. Hatsuko's facial features resembled mine rather than my husband's. Eyelids that were clearly double-folded, a slightly upturned nose, and hair and skin color were all the same as mine. I knew of this before, but these resemblances began to bear a certain meaning because of this small incident. I felt that there was a miniature me beside me.

Sometime later, my sister-in-law came to visit me. I used to feel unpleasant over her use of the fact that she had married into a big-city family and was living an urbane life as a weapon to compete with me. On that day, too, some suggestive words soon came out of her before Hatsuko and me.

"Sister-in-law, you look as relaxed and carefree as usual," said my sister-in-law, twisting her red-painted lips up to the right. It must have been one way of putting on an affected air, a way which she must have learned from somebody. I had noticed that each time I saw this sister-in-law after several months, some pretentious new mannerism was added to her expression or gesture or the way she talked, even while some of the previous ones disappeared without any trace, possibly having become old-fashioned.

"Do I look relaxed and carefree?" I laughed, feigning innocence. I have not lived in a carefree way; I have been managing this old house, which has been in the family for several generations, according to the instructions of my mother-in-law. At present, with my mother-in-law gone, everything in this house, all the old furniture and the shapeless old things, bears the stains of my fingerprints and of my breath. I am confident that by this time I have tamed every visible and invisible thing which I inherited from my mother-in-law.

"When I come back here sometimes, I feel one night is enough."

"Since mother is no longer here." Saying this I served the tea. Hatsuko was sitting sideways far away from the table, watching my sister-in-law with one eye.

"I don't mean that."

"You don't feel quite relaxed here?"

"No, just the opposite. I feel too much at home—so much so that I fear I would be left outside of the age." My sister-in-law looked around the room as if drawing a circle with her eyes.

"I do it in the way I like, since it's my house. Although I inherited it, it's my house now," I said aggressively.

"That's what it is. You lack imagination."

"For example?"

"Come to see my house once, Hatsuko." Sister-in-law smiled at Hatsuko. Hatsuko turned her face and then stretching her legs on the tatami mats, began exercising, touching her toes with her fingernails. On the wooden corridor beyond her a white moth was fluttering its wings.

"Above everything, it is the color. The walls are white, and the curtains on the windows are in bright colors. Do you, sister, have the courage to wear such original colors as yellow or green?"

I remained silent. When my sister-in-law started like that without restraint, I usually sank into silence. For me, the game is decided by ignoring the opponent, but she seemed to take my silence as my defeat. Such one-sided perceptions irritate me. I stood up, holding the box of cake which my sister-in-law had brought.

"You are just like my mother. The urbane style is to open the gift at once and enjoy it together with the guest. In this house, it is immediately put back in the kitchen closet. Before we know it, it starts growing mold."

At her words I sat down again. My sister-in-law began opening the package herself. I poured the tea again. Hatsuko, my sister-in-law, and I began eating the cake silently.

"What do you think of my eyebrows?" sister-in-law said suddenly.

"They look the same." In her made-up face, only the eyebrows were natural, forming a dark, fine-shaped bow.

"Aren't they too thick?"

"Isn't that fine?"

"I have not shaven; nor do I use pencils."

"It was the same with mother-in-law."

"I don't look like anyone else."

"On second thought, mother's eyebrows were not so distinct as yours," I was forced to say.

"Hatsuko-chan, what do you think?"

"About what?" Hatsuko opened her large eyes wide.

"Aren't your aunty's eyebrows too thick?"

"No, not at all."

"Is that right? I don't think so. You really don't think so?"

The moth which was fluttering its wings in the outside corridor had come into the room while I wasn't watching. The moth crawled on the mat, scratching its white body against it and scattering powder as its thick wings trembled.

"No matter what you say, they are too thick." Taking a small looking-glass out of her handbag, sister-in-law looked into it.

"Don't worry, aunt." Hatsuko's voice was cold.

"Are you saying my eyebrows are good ones?"

"It's true."

"Do you think so too, sister-in-law?"

"They are good eyebrows."

"No, no, they can't be."

I stood up, pretending that I was going to change the tea in the teapot. Hatsuko stood up too, following me. I intended to kick the moth away. I thought that by that action, this scene would be resolved. But Hatsuko did it before me. What my foot was trying to do, Hatsuko's foot did. The moth, having been kicked, fell outside on the open corridor. Hatsuko, putting on the wooden clog placed on the stepping stone, crushed the moth with the bottom of the clog.

That time, too, I should have felt proud of Hatsuko's response to my sister-in-law. Yet instead I felt the existence of a woman beside me, a woman who felt and acted exactly the same as I did. Because of this, my irritation over my sister-in-law disappeared, but a new irritation came over me.

It was a trifling thing. The first time it was Hatsuko's hand and the second time the action of her foot. But I felt as if the contents of my body flowed out in large quantities from its external structure while my hands and feet surely remained, or that my hands and feet were invaded by Hatsuko's hands and feet. This feeling clung to me. There are feelings which cling to you no matter how you try to shake them off. Under these circumstances, Hatsuko's behavior began to attract my attention further.

Hatsuko was subscribing to a series of novels for young girls. They were sent by parcel post, one volume each month. With nervous, sensitive movements of her hands, Hatsuko used to try to untie the knots in the thin linen string which bound the package over the wrapping paper but always she became so impatient that she cut them with scissors. The extremely methodical manner with which she started would invariably break down all of a sudden into a negligent manner. While watching her I was reminded of my own way when I tried to open packages. Hatsuko liked to sharpen pencils with a knife. She patiently and neatly shaved off the wooden part so that the wood would show a long cut surface, and then she sharpened the lead as fine as possible. She threw away without hesitation the pencils which became short, leaving only the long, matched pencils placed neatly in her pencil case. I couldn't help noticing that they looked surprisingly like the pencils in my own pencil case, which I kept nearby in order to keep the family books or my diary. I observed too her habit of throwing into the washing machine the handkerchief she had taken with her whether or not it was dirty, her quick motion to occupy an empty seat ahead of other passengers whenever she got on a train or bus, and the way she would flick a moth repeatedly with her finger until it landed on the open corridor when she found one on the tatami mats instead of crushing it with her fingers.

There were several differences, but the similarities particularly attracted my attention. As if Hatsuko were made of pebbles and iron powder and only the iron powder was drawn to the magnet held in my hand, the details of that part of her which resembled me gathered to my consciousness like distinct black spots.

So far as my visible habits are concerned, it could be that Hatsuko picked them up while watching me without intentionally imitating me. One day, however, I was forced to realize that that was not always the case. It was when I told Hatsuko to bring a large flower vase to the *tokonoma* alcove of the guest room, after having arranged the flowers myself. I followed Hatsuko, worried that the vase might have been too heavy for her. Hatsuko was placing the vase in the alcove, bending at the waist. Her shoulders looked square and hard, retaining the tension of carrying a heavy thing. She soon came back across the room. But when she approached the doorsill, she hesitated somewhat anxiously and looked back toward the alcove. Then returning to it again, she touched the vase with both her hands. It was not to correct its position, but she was touching it as if she wanted to confirm its existence with her hands. Then she came back.

"What were you doing just now?" I asked her at the doorsill.

"What do you mean?" Hatsuko looked up, wondering.

"Why did you go back to touch the vase again?"

"That is—" Hatsuko said smiling. Hatsuko laughed.

"Are you imitating me?"

"Do you do that too mother?" Hatsuko asked in return.

"No," I said in a hard voice.

"You see, I carried a heavy thing, then put it down and came back. I felt my hands were empty."

"So?"

"I don't know for sure, but—"

"Do you mean to say that if you touch it again then your hands relax?"

"You know, mother."

"How would I know?" I negated her again. I felt secretly that Hatsuko had begun to follow me even in such perceptions. During this period when I was regarding Hatsuko with suspicion, an incident involving a broken bone took place in Hatsuko's school class. It was a day when a PTA meeting was scheduled, just before the start of summer vacation. I was waiting near the stairway of the second floor, where I had gone together with a few others who had also come early. The voices of the students who had stayed on after class were ascending in a melancholy reverberation, so dispersed as to envelop the school buildings and yard together. A young woman instructor who was popular among the girl students, including Hatsuko, although she was not Hatsuko's classroom teacher, passed the waiting parents and went down the stairs. Then there was a loud scream and seven or eight girl students came running down the corridor. Sayoko was in front of these girls, and Hatsuko was visible among them. It seemed that they were competing to catch up to the teacher for some reason I did not know. The fierce group rushed down the stairway like an avalanche. The floor planks squeaked and the dust which flew up made the rays of the summer afternoon sun, slanting in from the window, feel thicker still. Just then a lame girl was climbing up the stairs, bending her leg step by step. She became caught up in the group which was rushing down. While looking down from the banister of the stairway at the movements of the bodies which became entangled with each other, I caught sight of a leg with a bandage on it hooked around the leg of Sayoko, who was trying to be first. Sayoko fell down, turning twice and thrice, and animal-like screams of surprise arose from the girl students. Sayoko lay with a face as flat as paper at the foot of the stairs.

"That girl did it," a high-pitched voice said, and it was Hatsuko. She was pointing to the lame girl who stood in the middle of the stairway with a worried look. At that moment I saw that there was a bandage on Hatsuko's slim ankle—perhaps she had sprained it at school that day. Something like a midwinter chill ran over my spine.

"It happened because that girl blocked us." Hatsuko, her eyes shining unnaturally, continued to look up at the lame girl.

"You bumped into her, is that right?" The young woman teacher went up the stairs two or three steps toward the lame girl.

"Uun." The deep, frightened eyes of the lame girl darkened.

"What is 'uun'? Answer in proper language."

The lame girl, without answering, looked vaguely around. The woman teacher raised her voice, looking around her.

"Everybody, what's the proper word for 'uun'?" Gay laughter arose from the girls. A nurse came and carried Sayoko in pain on her back to the school clinic. The woman teacher had not released the lame girl.

"I've told you to hold on to the banister when you use the stairway."

"I was holding on."

"Then this should not have happened."

"I haven't done anything."

"Your leg did it." Again the loud laughter of derision arose from the students.

I no longer felt like attending the PTA meeting which was about to start. The blood drained from my head and I was captured by a feeling that the back of my head was about to float away, turning to vapor.

A few days later we went to the beach. It was my husband who planned a two nights' trip for swimming at the beach for Masao and Hatsuko, who had only had the experience of swimming in the river. We arrived at the beach before noon and soon my husband and the children went swimming, leaving me alone. In the afternoon, all four of us had a boat ride. As my husband, who had belonged to the boat club in college, rowed us quite far out to sea, the land became blurred like a thin line drawn by pencil, and I felt that we were alone on the elevated surface of the sea, just the four of us. In front of me, the thick back of my husband, already showing a pinkish touch of sunburn, was moving regularly back and forth with the movement of his oars. The reflection of the sea's surface was like the glitter of thin steel plates, coming up one after another from the bottom, crushed before my eyes and dispersing into the powders of light. Even the voices of Masao and Hatsuko reverberated, glistening.

"Don't move around so much."

"I'm bored being still."

"Since we're going around the island, it will take another twenty minutes at least."

"I'll go to mother."

"You can't stand in the boat."

"I don't care."

"I wouldn't know even if you fell into the water."

"I can swim."

"You don't know—there may be sharks."

"Is that true?" Hatsuko's voice glided toward me. I raised my face from the surface of the sea. Hatsuko was facing my husband.

"Of course it's true," he answered with laughter in his voice.

"Have you ever seen one?"

"Father saw them in the South Seas when he was in the war." Masao drew in the air the long, sharp shape of a shark with his hands.

"They won't be here."

"When Hatsuko falls in, they will come from the South Seas," my husband teased her, joining Masao.

Before me there was a picture of happiness framed by the melting lights of the sky and sea. Since it was before me, it did not include me. What kept me away from it was Hatsuko. Over my husband's left shoulder, there was Masao in navy blue swimming trunks, and over his right shoulder was Hatsuko in a black bathing suit. I realized that I was placed by accident in a position in which I faced Hatsuko. There was no way to change my position for about thirty minutes, until the boat returned to the beach.

As if looking at the image of myself in a large mirror placed before me, I, sitting in the back, gazed at Hatsuko. Her appearance resembled mine, her habits resembled mine, her feelings resembled mine— That was all still nothing. I had been hurt deeply by what Hatsuko had done a few days before. My memory of my own experience of long ago corresponded to it completely. On that sea filled with lights, hearing the voices of the children and my husband with no traces of shadow, it was not pleasant to be reminded so vividly of things that had happened more than ten years ago.

My class teacher was an old man who was disliked by the students because he was cunning. He did not have his own family, and he often volunteered to sleep over at school as a night watchman. He kept a goldfish in a large fishbowl in the room kept for the night watchman. People said that between classes he used to go to the room and talk to the goldfish, saying, "Goldfish, how are you?" One day after school, I offered uncalled for service; together with Yoshiko I volunteered to clean the fishbowl. Although the old man had changed the water every day, the exterior of the bowl was a greasy cloud from being handled. Yoshiko was a classmate who was slighted by her friends but who always followed me. With the sound of his pleading voice asking us repeatedly to be careful at our backs, we went out into the schoolyard, holding the bowl on both sides. I saw the face of the old man, who was obviously anxious, looking out the window. It was very difficult to carry the large bowl, for not only was it heavy but also the water swayed thickly inside. The more tense I became, the more uncertain my hands felt, which frightened even me. I felt the distance to the water faucet at the washing place to be extremely great. The persistent echo of the old man telling us to be careful rang in my ears. The smell of the moss of the courtyard where the sunshine did not reach and the vinegary smell of Yoshiko's breath struck my nose especially strongly. I suddenly felt tired. One shoe slipped on the damp moss, and for a moment I felt that the bowl, which slipped from my hands, was suspended in midair. Immediately the sound of glass crashing was heard at my feet and the water exploded. The old man's voice, which could have been either a scream or a cry of condemnation, could be heard from the window at Yoshiko's back. The red belly of the goldfish was trembling among the broken pieces of glass. When the old man appeared, panting, unexpected words came out with a smoothness which could only be called spontaneous. Yoshiko, what did you do? Because you let go, Yoshiko, it is your fault. Yoshiko at that time looked up at the old man with a humiliated glance— I don't know why. Bending forward, the old man struck her head. Yoshiko fell down and fainted, hitting the back of her head on the concrete edge of the courtyard.

When my husband's back straightened with his rowing, I could see only the upper part of Hatsuko, and when it bent forward, I could see her entire figure, with her slender waist wrapped tightly in the black bathing suit. Both the suit and her hair had dried in the boat. Some of the sand which stuck to her arms when she lay on the beach remained on her right arm. The same blood as mine was running through that body, I felt with a strong, sure feeling which I had not felt before.

Suddenly Hatsuko stood up. The boat tipped sharply to the right as Hatsuko stretched both hands up in the air and fell into the sea. She choked with the salt water, but other than that she was not hurt. When Masao said,

"I told you not to stand up, didn't I?"

Hatsuko replied laughing,

"There weren't any sharks."

I looked far away, taking my eyes off Hatsuko. Stretched waves marked the few stripes on the surface of the sea. They were constantly moving, but as a whole the same striped shapes remained all the time. As I looked at it vacantly, the sea appeared immobile, as if it were a shining steel sheet. The vision of a shark springing up to break that surface crossed my sight bewitchingly. I could see before my eyes a vision of Hatsuko's body, swallowed by its sharp, wide-open mouth, shining more vividly red than in reality.

"I who was disliked by mother without knowing why," Hatsuko had written. She could not have known the reason, for I made it my task to hide it from her. Since Hatsuko was sensitive like me, I had to be perfect in hiding it from her. Poor Hatsuko. But I was trying as hard as I could. As hard as I could? Trying hard not to climb up the numbers of plus but to run down the numbers of minus—what does it mean to try hard for something which leaves only a sterile wasteland inside me?

In the early years of my marriage, when I had just come to the house, I used to be aware of its peculiar smell. After living here for some years, I was not conscious of it every day, but when I returned from outside or when I stood up suddenly after being immersed in some thought, the smell struck my nostrils with unexpected vividness. Since the smell was particularly strong in a room closed for some time or in the storage room, it must have been the smell of mold on the old tatami, paper doors, and other fixtures. It was a smell which had become ingrained in them over a long period of time, and it made me think that the generation preceding me and the generation before that must have lived and died smelling the same smell. When I came to the house, my sister-in-law was already married and only my mother-in-law and my husband were living here. I felt that I had brought the smell of a young woman into this house. After my mother-in-law died, my own thick, sweet body smell and my husband's body smell mixed with the smell of cigarettes and pomade were the only human smells I detected among the thick layers of moldy smell. Outside of my own, the only other human smell I encountered was that of my husband, but when Masao entered high school, I realized that another human smell had been added. It was a somewhat oily, sticky, animal-like smell, stinging the nose sharply. I encountered the smell of my husband and of Masao several times a day; it arose especially strongly not only when I passed them in the corridors, but also when I picked up their pajamas or pillowcases. In this way I felt the existence of my husband and Masao in the house.

One summer morning, I think it was when Hatsuko was in the third year of junior high school, I was sitting before the mirror to arrange my hair. It was soon after I sent my husband, Masao, and Hatsuko off. Usually I get up first and after washing up and fixing my hair, I give them breakfast and send them off. On that day, however, I got up later than usual, so I sat down before the mirror after they left. The minute I sat down I felt that the cushion, usually cold-feeling, was warm and a sweet smell, seemingly mine, lingered there. It had to be the smell of Hatsuko, who must have sat there before she left. While I was feeling that Hatsuko had come to resemble me even in her smell, I heard her footsteps running back on the stone pavement, and Hatsuko called me in a high-pitched voice. Thinking of the smell, I came to the entrance wondering if she had started her first menstruation. But Hatsuko, saying that she had forgotten her bus pass, ran upstairs to her room.

"You are not hiding it, are you? You must tell mother when you have menstruation," I said to Hatsuko, who had come down the stairs. Hatsuko immediately dismissed my overserious tone with dry laughter.

"Don't worry. It will come late with me."

"How do you know it will come late with you?"

"You said it was late with you."

"How can you say that it will be late with you if it was late with me? How do you know it will be the same with you and me?" I said in a strong tone, feeling I was being too persistent even for me.

"I'll miss the bus." Hatsuko turned and ran out on the stone pavement. I watched her disappear outside the gate, her shoulder-length hair, which had the same color as mine, swaying, and then I returned to the mirror.

I picked up a strand of hair which had fallen on the cushion. Then I plucked one strand of my own. Raising the two before my eyes, I looked at them against the light. The black hairs had a bluish tint, and only the shine of the blue appeared when they were looked at against the light. The two were indistinguishable. Taking a box of matches from the kitchen, I went out to the open corridor. Holding the two strands slightly separated between my thumb and index finger, I set fire to them from the bottom. I did so because I recalled that during the war I used to distinguish silk fiber from rayon staple fiber, which looked the same, by examining the way they shrank and their smell when burned. The two hairs shrank up from the bottom, forming the same shape, making a sound similar to thin sticks of burning fireworks and shrinking into black dots similar to the remains of fireworks. Twice, three times I brought the match flame to the hairs. They soon disappeared from the bottom, leaving only two coarse, black balls almost touching my fingers. "What on earth are you doing?" some part of my consciousness asked me coldly.

A month later, Hatsuko had her first period. I came to smell still more strongly and distinctly the same odor similar to mine that I had smelled before the mirror that day. It could be smelled at different places and different times—on the living room tatami, on the wall of the corridor, in the corner of the bathroom, near the chest of drawers at the foot of the stairway, and in Hatsuko's room. Previously, the smell of others which I encountered in this house were the body smell of my husband mixed with the smells of cigarettes and pomade, and the body smell of Masao, a sticky, sharp stimulating smell particular to boys in puberty. My own body smell stayed with me always and went wherever I went. Yet with Hatsuko's growth, I was forced to smell my body smell even in places where I had not gone. I was forced to have the strange feeling of encountering my body smell outside of myself.

Hatsuko's body became rounder. The skin on her arms, legs, neck, and cheeks began to take on a vivid shine. There were times when I caught sight of her back when she came out of the bath and I was surprised by her well-developed muscles. Her waist was deeply curved and her buttocks were extended, making a slight curve. Hatsuko suddenly became feminine and like me.

One day in early autumn I was staying in bed, having caught a cold. I was running a slight fever and my body felt heavy and swollen, as if it contained lots of water. I put a mattress in the former maid's room near the entrance so that I could get up easily when someone came to the house. After entering the figures in the home account book while lying on my stomach, I looked vacantly at the wood designs of the ceiling plates, which were tarnished black. The *shōji* screens were closed to prevent the western sun from coming into the room. Since I rarely stayed in bed during the day, usually working hard inside and outside the house, I realized for the first time that the light which came through the *shōji* screen was curiously bright. It was a brightness one might call luxurious or abstract, a uniform brightness diffusing without variations in the intensity of brightness. I felt it inducing sleep just as darkness does, and soon my consciousness spread itself as wide as the room, floating in the brightness. It swayed half-asleep at the border.

"Ohh, that's all right."

A bouncing voice could be heard. The impression that it came from outside immediately faded into an impression that it came from inside of me, and I wondered vacantly about the overlapping of the two.

"These are different from ordinary ones. Once you try them you'll be able to tell right away." I heard the voice of a peasant woman who had come from far away a few days before to sell chestnuts, a voice that hissed with the breath leaking through her teeth. I realized that the conversation I had exchanged with the peasant woman at the entrance was being repeated in my dream.

"On our mountain we have lots of them. We don't have to buy them," I was saying in my dream.

A smile which could be understood either as good-natured or cunning spread across the peasant woman's face. She had a gesture of rubbing her right hand against her traditional, navy blue cotton slacks. She repeated this two or three times without any apparent meaning.

"But young lady, I'll make it real cheap. One hundred yen for all these. It is not the usual price."

My consciousness, which was spread thin, began sinking. I began seeing the light come through the screen, and the wood design of the ceiling plate appeared vividly again. But the voices in conversation continued at the entrance.

"I told you we really don't need them. Whether they are cheap or expensive, we have too many—we have to throw them away. Ha, ha."

The voice in the dream which I thought mine was the voice of Hatsuko. The slightly high-pitched voice and the already grown-up tone were the same as mine. The conversation stopped. I heard the sound of the peasant woman's rubber slippers batting on the stone pavement gradually fading, and then Hatsuko's spirited footsteps. After that a vacuum-like stillness spread through the house. How eerie, I thought in bed, that while I am lying here like this another me is walking around me and talking to the peasant woman.

Still lying down, I heard a voice from the woods and turned my head toward an eastern window. There I could see the slope which climbs up from the house to the woods. The setting sun cast its rays there, and the space cut out by the small window looked curiously beautiful. The yellowish maple leaves appeared to have been dyed with a particularly vivid yellow. Soon Masao entered into this space. He had a ball in his hand. Then Hatsuko appeared. It seemed they had gone to get the ball which they had lost in the woods some days before. Hatsuko bent forward as if she had stumbled on something. Her mouth was shaped as if she were shouting something to Masao, who was no longer within my sight. Then she bent down on the slope. But soon she stood up, holding one sneaker in her hand. Turning it upside down, she shook it with a swift movement of her hand. Maybe pebbles had gotten into it. She then bent down again, and standing up after putting on her sneaker, stood in the same posture quietly facing the setting sun. In the space which was almost too bright—as if it were a screen singled out by spotlights—Hatsuko's figure was striking and vigorous. Red sweater, brown and gray checkered skirt, white, almost translucent cheeks, bluish black hair—it seemed as if the colors of her clothes were brought out not by the setting sun but by the life inside her.

The book of home economy into which I was entering the household expenses was lying next to the pillow; I stretched my hand to it and picked up the pencil. I raised the upper half of my body from the mattress. The pencil was cut so as to show the long wooden surface and the lead was thin and sharply pointed. Holding the middle part between my thumb and index finger, I placed the end of the pencil before my right eye. Closing my left eye, I aimed so that the tip of the pencil was pointed toward Hatsuko. I aimed the gun, so to speak. Hatsuko was still standing in the brightness of the setting sun, looking in the air somewhere far away, showing her profile to me. Hatsuko, move away quickly. If you don't, mother will shoot you. Hatsuko did not move. I held the pencil still. Quickly retreat to some place where mother cannot see you. The pencil in my hand felt heavy and hateful. When Hatsuko's figure moved slowly from my sight and disappeared, I felt relieved and at the same time tired.

One day on my way back from shopping I met Hatsuko returning from school. When we came to the bridge after passing through the shopping district, a chilly wind, coming from the upper river in wide waves, blew against our cheeks. Blackish-purple ripples formed on the river as if numerous, jostling sardines were cutting the surface with their dorsal fins. Across the bridge, the old residential district extended slowly up the hill. I asked Hatsuko about school and she talked about her friends. It had become my custom to ask my husband, Masao, and

Hatsuko about their day and to listen attentively to everything they said, even though it did not necessarily interest me. When Hatsuko and I were coming back together talking about nothing in particular, an old woman with gray hair appeared in front of us. She was coming toward us at a rapid pace, hitting the ground quickly with a cane.

The old woman came on, looking straight at me and Hatsuko. She had an air of arrogance which indicated she did not consider such a gaze to be impolite.

"Ho, ho." The old woman stopped right in front of us, making a sound like a sigh.

"Are you mother and daughter?" Her round, wide-open eyes looked crossed only when she spoke.

"You know us, don't you? I am Matsuyama, the house with the *karatachi* hedge." As I hesitated to talk to the old woman directly, I spoke in a tone meant to cut the conversation short. She was living separately from her family in a dilapidated house at the foot of the mountain. She was called a crazy old woman, for she frequently upset people by saying strange things.

"Is that right?" So saying, the old woman started walking again. As she passed, I caught the sound of her throat choked with mucus.

"Ho, ho."

Again I heard the voice at my back. When I turned back, the old woman had stopped and was looking at us.

"The figures are the same even from the back."

"Because she is my daughter," I said, feeling bitter.

"Yes, yes." The old woman smiled. While her eyes did not smile at all, there was something which emerged from the spaces among the wrinkles which creased her entire face that I took for a smile. The old woman disappeared around the corner of the residential district, the hem of her dark clothes fluttering.

Half a year after that I met the old woman again. It was a damp, cold day in early spring, when the bare branches of the deciduous trees appeared white like bones. Coming out from between the woods and a white wall, the old woman appeared unexpectedly and approached me with her unsmiling smile.

"What happened to your daughter today?" she said, striking the ground sharply with the tip of her cane.

"I am not always with her."

"Yes, yes. That's best. It will be better for you to walk alone."

She was annoying me and I tried to pass her, but she continued to pursue me.

"It's better that two similar ones do not stay together. It's inviting trouble." Suddenly a scream which might have torn the sky came out of her mouth. I stepped back, stunned. The old woman stooped on the ground. She began trembling, bending her back like a ball. When the trembling ceased, the voice came out elongated, turning from low to high. After this was repeated three or four times, the old woman started to cough hard. It must have been a fit of asthma. I felt that that ugly, shrunken body contained an ominous power to see through the mind of others. After wondering for some time whether I should help her, I fled, afraid of having anything to do with that power.

One summer evening I sat down before the mirror after hurriedly washing off the perspiration in the bath and changing into a clean, well-starched *yukata.* The color of blood comes to the surface of the skin after a bath. Although I was past forty, my skin did not show much decline. I took out of a drawer a lipstick I did not usually use. The metal of the case was rusted here and there, and the lipstick inside was whitish on the surface as if some mold had grown there. I put it on my lips carefully and thick. My face came to look still fairer because of the red. The sweet smell of the perfume of the lipstick induced a gay feeling in me. I tried smiling. I saw the gay face of a woman in the mirror. It was another face buried under the dark layers and layers of life. I thought I could have lived with such a face. My sister-in-law was a woman who had chosen such a life. But I did not do so. I never regretted it at all. Yet I did think about the gorgeous woman which had been crushed inside of me. If given a chance it could have bloomed into a large flower spreading wide its pink petals and wafting around a sweet fragrance. Such a flower which could not bloom existed inside of me. It existed inside of me without shrinking or withering, no, containing a still richer fragrance precisely because it could not bloom fully.

Standing up before the mirror, I walked toward the guest room. I met Hatsuko coming down from the second floor. Seeing that she was wearing a cream-colored dress for going out, I asked her,

"Where are you going?"

"I'm not going anywhere."

"What's that dress for?"

"Where are you going yourself, Mother?"

I realized that although I had applied the lipstick meaning to wipe it off immediately, I had come out carelessly forgetting to do to.

"Since we have a guest—" I said, quite disconcerted.

I returned to the mirror and wiped off my lipstick completely. In the guest room, my husband was playing chess with the guest. Our guest was the young master of the bakery across the river, the most famous bakery in this small city. He left the business in the hands of the employees and spent his days fishing and collecting antique art. Here is *ayu,* freshly caught. I brought it for you *okusan.* He used to bring sweetfish every year when the ban on fishing sweetfish was removed, saying here is sweetfish, I brought it for you madam. He was four or five years younger than I, but looked even younger. Although he was nonchalant, he was quite sensitive, and a look that surfaced sometimes when he smiled induced a coquettish feeling in me. It was because of this guest that I had thought of putting on lipstick.

Seeing that my face had returned to its usual hard state after I wiped off the lipstick, I stood up saying "all right" to myself. I went to the living room to fetch a fan. There were only three fans, although four were usually piled there. These fans were commonplace ones with pictures of evening primroses, bellflowers, and daisies, and they were used by my husband, Masao, and Hatsuko. I could not find a large fan on which a peacock was drawn with golden powder on a red background. It was my own fan which a friend who was studying painting had made for me. Taking therefore the fan with the bellflowers drawing, I went to the guest room.

I stopped short at the doorway, surprised. Hatsuko, sitting beside the guest and my husband, who were playing chess silently, was fluttering that very fan. She was making a breeze not for herself but for the guest.

"Uncle, it is humid, isn't it? The breeze stops coming into this room in the evening, although I don't know why," Hatsuko was saying.

"It will be bearable if you two fan for us," my husband said, looking up at me still standing at the doorway.

Instead of taking the cue, I returned to the living room to put back the fan with the bellflowers drawing. Bringing a soda bottle and two glasses from the kitchen on a tray, I returned to the guest room. When I put the tray down, Hatsuko's hands reached out quietly, opened the bottle, and began pouring soda into the glasses.

"Please." She extended the tray to the side of the chessboard with the gesture of a mistress. My role usurped, I was left with my hands dangling helplessly in the air. The sand pebbles of irritation started making noise inside of me. I went out to the open corridor silently. The surface of the wooden planks, exposed to the sun for long years, felt coarse and sandy to the soles of my feet. A purplish line of smoke made by mosquito-repellent incense was rising at the end of the corridor. For some reason its smell, which used to cling to my nose, did not strike me at all.

On hearing Hatsuko's voice, I turned back to the room. My husband was trying to catch a cup which had turned over. He used to upset cups when he was absorbed in playing chess. Hatsuko was wiping the hem of her one-piece dress with a handkerchief with a hitting motion.

"It's because you are wearing such a dress," I said strictly and left the room in order to fetch a wiping cloth. My eyes met Hatsuko's. Hatsuko moved her eyes to the guest.

"Uncle, you met me at the bridge some time ago when I was wearing this dress. You said 'Hatsuko-san, you look pretty,' didn't you?"

"Is that right?" The guest smiled ambiguously and looked at me.

"Will you go to uncle as a bride, how about it?" my husband said, making a hitting noise as he put down a chess piece on a square on the board.

Hatsuko laughed with a lovely, chuckling voice. She started fanning herself. The fluttering red background and golden design made either the red or the gold more vivid than the other according to the angle of the fan. Hatsuko's cheeks reflecting the colors, looked charming.

"We have an old electric fan but no one likes it. The breeze from a hand fan is softer," I heard Hatsuko say at my back in exactly the same tone as mine.

I remained immobile, in the kitchen, holding the wiping cloth and gazing into space. Hatsuko's behavior was completely unintentional. Yet all the more for that reason the erotic atmosphere of which she herself was unaware flickered like a thin veil. What a ridiculous triangular relationship! I recognized in Hatsuko the charming woman with lipstick I had seen in the mirror. The only difference was the one-piece dress instead of a *yukata;* Hatsuko was acting the part of the woman I had seen momentarily in my dream. This is what irritated me. It would have been nothing if a completely different woman had been there. But Hatsuko had stolen from me the woman whom although longing for I had locked up, the woman who applied lipstick but later wiped it off.

A few days later the young master of the Japanese bakery came wearing a straw hat to take Hatsuko fishing. The incident did not disturb me much at the time. I sent Hatsuko out in light clothes, letting her wear sneakers so that it would be easy to walk on the slippery slopes. When I sat down vacantly in the living room, another scene glided into my vision with unexpected vividness. From that moment I started seeing—

Hatsuko and the man are going down the gradual slope in the old residential district. Dry dust flies up, clinging to their perspiring legs. Turning at the temple, the road continues alongside the yellowed low clay wall of the temple. As they come out to the paved road, the air expands free and wide, and a bleached brightness floats on the river surface under the summer sun. The two sit down on the romance seat in a bus. Going in the direction of uptown on the road along the river, Hatsuko inhales the smell of the river breeze. Yet the body smell of the man next to her is retained in the bottom of the river breeze blowing through. Hatsuko is puzzled by the mingling of the two smells. The river gradually becomes narrower and deeply curved, and the dampness and shade of the mountain creep along its surface. Suddenly a strong gust of wind blows in through the window and the man's straw hat is almost blown off. Oh uncle. Hatsuko swiftly catches the hat floating in the air. The man laughs. Because of this little incident the intimacy between them increases. Getting off at the mountain and walking down the slope, they come to the riverbank. The pebbles at the riverbank could be felt through the soles of the sneakers. On the near side of the river, layers of shining clear water meander along, but at the far bank a cliff juts up and the blue-black water at its foot is stagnant. Hatsuko gazes at the disparity between this darkness and light. The man walks further upstream. This side of the river too becomes narrower and the slope rocky. They move carefully, feeling for the indented surface of the rock. The man stretches his hand out to her. I am all right uncle. Hatsuko withdraws her hands. At that moment, she slips on the wet, sloping rock. The man quickly grabs her arm. But the lower part of Hatsuko's body has already fallen into the water. I remain still, for I feel it pleasant to have my body in the water. I look up at his eyes while he holds my arm. The warmth of his body comes creeping through my arm. What happened? Hatsuko shakes her head. There is no way to answer. I do not want to move. The warmth of the man's body transmitted through my arm faintly lights the candle of my body. My body begins melting, turning into wax. The river flows from the lower part of my body. The river gradually becomes warm and my body and the river become a continuous, long flow of wax floating, melting down the river. The man is smiling. The tempting, smiling gaze which I used to recognize pierces my eyes. But the man is smiling as if unaware of my response. My body is melting drop by drop, and since it continues flowing endlessly down the river, it stretches endlessly without knowing to which sea it will flow. At the same time, it continues to flow warm and helplessly pleasant. Hatsuko-san what happened? Hatsuko smiles dreamily.

Regaining myself, I stood up in the living room. It was not me who went with the man. It was Hatsuko who went with him, leaving me. I went out the entranceway and started walking quickly on the stone pavement. But it was too late. Hatsuko and the man were already far away and I had no way of taking back my self which was taken away by Hatsuko.

Just as I came out of the gate, that old woman was passing before it. She seemed to have said something to me, but I could not catch it. Her eyes, set as if crossed, gazed at me steadily. Those ominous eyes which pierce others' minds. Turning her back swiftly with a sigh, she moved on, striking the ground quickly with her cane. I chased her. No, I did not. I felt that the old woman had read my mind, and I retreated behind the gate immediately.

It was in my dream that night that I did chase her. The figure of the old woman walking with the cane in her accustomed manner made her look from the back like she was walking with three legs. The thick fumes of darkness swallowed the old woman, and soon I lost sight of her. The darkness was thick and sticky, like an oily fume. When I reached the foot of the mountain, waving my arms as if swimming in order to wipe away the oily fume of darkness, I found a large, dilapidated house, still darker than the darkness. Entering by the back door, I stood at the thatched annex where the old woman was living alone, away from her family. Although I called out, it did not seem that anyone was there. When I raised my eyes after sitting down in the open corridor, I could see numerous birds perched on a huge tree whose leaves had fallen. Curiously spotlighted, only that space was bright, and the birds stayed immobile without crying or flapping their wings.

"I have been waiting for you." The old woman appeared, something that looked like a smile floating out from among the wrinkles. She carried a blackened, middle-sized box. Instead of coming close to me and sitting in the open corridor, she sat at the edge of the room on the tatami mats. Untying the purple string with tassels, she took an old copper mirror out of the box.

"Look into this." She extended it to me.

"What is this?" I took the heavy, cold mirror in my hand.

"It's my eyes. Reflect your face in my eyes." Hesitating, I looked at the old woman. Her gray hair hung down, dry like straw. Retaining no moisture, her skin was like leather.

"You don't know. Look into this mirror," the old woman insisted stickily.

I looked into the mirror. My face was reflected in vague outline on the dull surface. While I stared at it the face began gradually to change. Or I should say that from behind the blue-brown surface of the mirror a strange, unfamiliar face emerged vaguely, and it overlapped my face. The face which revealed itself contained anger.

"It is the face of mother." At these words of the old woman, the image on the mirror's surface disappeared.

"It was not my mother's," I said. Then the copper mirror in my hand disappeared too, and looking up I saw that the old woman's eyes were gazing at me with the same color as the mirror's surface.

"It is the face of mother itself in general." The old woman laughed with a husky voice. Holding her right hand straight as if it were a blade, she hit her left wrist with it. The artery was cut and the blood began spurting out. The spray of blood made a veil between the old woman and me, and she began to fade away, only her despising laughter resounding; the blood crept stickily over the tatami mats, the flow seeming to displace the darkness, and the voice of the old woman could be heard from far away saying it was the blood of women, and the sky was filled with a red-black sticky secretion, and the old woman stretched out her hands and scooped it up saying it was the blood of women, look there is a limitless amount, scooped and scooped and it was still there; it is transmitted to the woman who comes out of your stomach, then to another woman who comes out of that woman, and what is transmitted is woman's karma, here try to scoop it, where can you find maternal love? It is nothing but an illusion manufactured by men. Look, look, there, there is only blood, why is there such a thing?

"On October twenty-fourth, I will visit you after a long absence," Hatsuko wrote in her letter. I felt from the morning something restlessly swaying in the corner of my mind. But it could not be something significant after all these years. At breakfast, my husband spoke only about his first grandchild, whom he was going to see for the first time, and Masao said that he would buy cakes which Hatsuko used to like at the Matsukawaya Western bakery on his way home. You can get French cakes anywhere nowadays I said sternly. After the two of them left for work, and I finished cleaning and shopping, I felt there was nothing to do. The time in the afternoon which was usually occupied by constant working became empty as though waiting for Hatsuko to glide into it. She had not mentioned when she would arrive. Taking the train schedule from the drawer, I tried to see what time she would arrive if she took a certain train and what time if another. But since the schedule was an old one, it was not very reliable. I was worried about facing Hatsuko one to one if she arrived before my husband and Masao came home.

It was a cloudy day with soft sunshine pouring down vacantly. To the persimmon trees, all of whose leaves had fallen off because of the typhoon a few days before, only the red persimmon fruit was clinging, and it looked as if numerous agate hair ornaments were stuck against the sky.

I changed into a kimono which had red stripes on gray cloth in order to welcome Hatsuko. Nowadays I usually wear dresses in both summer and winter. The reason I deliberately wore a kimono was that I did not want to look like Hatsuko. I laughed at myself for still being concerned about such a thing. After sending Hatsuko to a far-off college, thus pushing her to a place where she could not reach me, my emotions dried up, and I became coldly bright like the dried-up bottom of a swamp where the dirt remains. Hatsuko did not come home while in college except for New Year's day, not even during the summer and spring vacations, saying that she was busy working part-time. She found a job there and married, finding her husband by herself. Now in what manner shall I receive her?

I thought of picking persimmons to kill time. When I climbed up the ladder halfway to the top of the tree, an expanse of sky opened dimly. The cloudy heavens became dyed red around the western mountains, the color becoming thinner as it seeped gradually toward the middle of the sky. It looked as if the sky had a deep wound at one end from which blood was seeping to cover the whole sky. Cutting them with scissors, I threw persimmons one after another into a bamboo basket hanging on a branch. The evening wind blew against my cheeks with a chilly pleasantness.

The soft sound of footsteps made by sandals walking on the pebbles of the stone pavement came from afar. Still on the ladder, I stretched my neck to look over the *karatachi* hedge. I could not believe it, but the figure in a kimono that appeared every so often among the trees must be Hatsuko's. While picking the persimmons, I was expecting to hear the sound of the shoes of Hatsuko clad in a dress. Hatsuko appeared distinctly. She came walking without noticing me in the tree. A brown kimono, possibly made of *meisen* silk, a very red *obi* sash, a pink tie on her *obi,* holding a child with a cap in her left arm and a large satchel bag in her right hand. I opened my awakened eyes feeling as if I were watching myself in the past. I must have looked like that when over twenty years ago I walked this road holding Hatsuko.

Hatsuko raised her eyes and recognized me. Although she did not smile, her facial expression changed quickly.

"Here, I'll throw a persimmon." In order to make the meeting gay, I tossed a persimmon gently at Hatsuko. Drawing a curved line, it fell through the air. Then Hatsuko, swiftly putting her satchel on the ground, caught the persimmon easily in one hand as it fell to eye-level while holding the child in the other hand.

"Am I not good?" Hatsuko laughed. That kind of spontaneous skillfulness was just like my own, but it did not evoke any feeling in me now. I came to the front of the gate.

"This is your grandmother." Hatsuko turned the child's face to me.

"I don't like to be called such."

"You think you are still young?"

"I have aged all right."

"Her name is Misako." Raising the knit cap, she showed the child's face.

"So it was a girl." So saying, I walked quickly on the pavement ahead of her. I would have felt much relieved if it had been a boy.

Filled with family joy, the supper, held in the guest room, lasted a long time. My husband already showed the uncontrollably happy face of a grandfather when he held his grandchild or held her hand while she walked. Hatsuko and Masao talked garrulously, as they had in the past. I acted according to their mood, but I felt I was in an air pocket in the joyous atmosphere of the family.

I went out to the open corridor alone. Darkness engulfed the garden. A corner of the sky looked milky perhaps because the moon was hidden behind the clouds. The coldness rose, creeping from the surface of the ground. Outside there was a stillness as if everything was holding its breath in preparation for hibernation, but behind me, in the guest room, the talk and laughter swelled, gay, noisy, and with the smell of life.

Hatsuko came out to the open corridor. She tried to extend her child to me.

"Take her in your arms once."

I was forced to take that heavy, damp, warm thing in my arms. Because I happened to bear Hatsuko, my blood ran in her, and since Hatsuko bore Misako, my blood continued to run even in Misako. I am not able to relax with Masao either. When Masao marries soon and has a daughter, I will be transmitted there and will be relived in her. In this way, from next to next, I will continue to expand limitlessly into the dark space of the future. The thought gave me an ominous feeling.

The child whines. Its carp-like mouth opened and I noticed four little teeth on top and four below. The vivid pink of the mouth's inner skin was exposed. I returned the child to Hatsuko's arms.

"She looks like me, doesn't she?" Hatsuko said. But there were no distinct features yet on the infant's fat face. I imagined future days when gradually from a certain time all of a sudden this face would come to resemble that of Hatsuko.

"You too bore a girl," I said, smiling thinly. I checked my impulse to say that it will begin with you now.

十二

The Smile of a Mountain Witch

Ohba Minako

I would like to tell you about a legendary witch who lives in the mountains. Her straggly gray hair tied with string, she waits there for a man from the village to lose his way, meaning to devour him. When an unknowing young man asks to be put up for the night, the owner of the house grins, a comb with teeth missing here and there clutched between her teeth. As he feels a cold chill run up and down his spine beholding this eerie hag of a woman, her yellowed teeth shining under the flickering lamp, she says, "You just thought 'What an uncanny woman she is! Like an old, monster cat!' didn't you?"

Startled, the young man thinks to himself, "Don't tell me she's planning to devour me in the middle of the night!"

Stealing a glance at her from under his brows, the man gulps down a bowl of millet porridge. Without a moment's hesitation she tells him, "You just thought in your mind, 'Don't tell me she's planning to devour me in the middle of the night!' didn't you!"

The man, turning pale, quickly replies, "I was just thinking that with this warm bowl of porridge I finally feel relaxed, and that my fatigue is catching up with me." But with his body as hard as ice, he thinks to himself, "The reason she's boiling such a big pot of water must be because she is preparing to cook me in it in the middle of the night!"

With a sly grin, the old witch says, "You just thought to yourself, 'The reason she's boiling such a big pot of water must be because she is

---

Yamauba no bishō (1976). Translated by Noriko Mizuta Lippit, assisted by Mariko Ochi, with the permission of the author.

preparing to cook me in it in the middle of the night!' didn't you!"

The man becomes even more terrified. "You accuse me wrongly—I was only thinking that I'm really tired from walking all day and that I ought to excuse myself and retire for the night while I'm still warm from the porridge, so that I may start early tomorrow morning."

But he thinks to himself, "What a spooky old hag! This monster cat of a woman must be one of those old witches who live up in the mountains I hear so much about. Or else she wouldn't read my mind so well!"

Without a moment's delay, the mountain witch says, "You just thought, 'What a spooky old hag! This monster cat of a woman must be one of those old witches who live up in the mountains I hear so much about. Or else she wouldn't read my mind so well!' "

The man becomes so frightened that he can hardly keep his teeth from chattering, but he manages to shuffle his body along on his shaking knees. He says, "Well, let me excuse myself and retire—"

Practically crawling into the next room, the man lays his body down on a straw mat without even undoing his traveling attire. The old witch follows him with a sidelong glance and says, "You're thinking to yourself now that you'll wait to find the slightest chance to escape."

Indeed, the man had lain down hoping to take her off her guard, so that he might find an opportunity to run away.

In any case, these old mountain witches are able to read a person's mind every time, and in the end the victim runs for his life away from her abode. The old witch pursues him, and the man just keeps running for his life. At least this is the form the classic mountain-witch tales assume.

But surely these old witches cannot have been wrinkled old hags from birth. At one time they must have been babies with skin like freshly pounded rice cakes and the faint, sweet-sour odor peculiar to the newborn. They must have been maidens seducing men with their moist, glossy complexions of polished silk. Their shining nails of tiny pink shells must have dug into the shoulders of men who suffocated in ecstasy between their lovers' plump breasts.

For one reason or another, however, we never hear about young witches living up in the mountains. It seems that the young ones cannot bear to remain in their hermitage, and their stories become transformed into stories of cranes, foxes, snowy herons, or other beasts or birds. They then become beautiful wives and live in human settlements.

These beasts that disguise themselves as human women invariably make extremely faithful spouses; they are very smart and full of delicate sentiments. Yet their fate somehow is inevitably tragic. Usually by the end of their tales they run back into the mountains, their fur or feathers pitifully fallen. Perhaps these poor creatures, with all their bitterness and resentment, turn into mountain witches. After all, devouring may be an expression of ultimate affection. Does not a mother in an emotional moment often squeeze her child and exclaim, "You're so dear to me I could eat you up!"?

Now, the woman about whom I am going to speak was a genuine mountain witch.

She died at the age of sixty-two.

At sixty-two, when her soulless body was cleansed with rubbing alcohol, her skin was bright and juvenescent like the wax figure of a goddess. Her hair was half-white, and on the mound at the end of her gently sloping belly were a few strands of silver. Yet around her calmly shut eyelids and her faintly smiling lips lingered a strange innocence and the bashfulness of a little girl who is forcing a smile even though she is about to burst out crying.

Indeed, she was the mountain witch of mountain witches. But even though she often longed for a hermitage on the mountains, she never lived in one, and she spent her entire life in the dwellings of a human settlement.

She had been a mountain witch ever since she could remember.

When she was still at a tender age and had not yet quite learned to use the bathroom, she would be so engrossed in play that she often had accidents. She would say to her mother who came running, "Oh you naughty girl. You've got to tell Mommy on time before it's too late. Oh dear, and today we don't have any change left for you—"

As her mother burst out laughing, she would go on, saying, "Really I'm no match for this child!— What can I say!"

At night, when her father was late coming home and her mother glanced at the clock on the wall, she would immediately say, "What in the world is he up to, coming home late night after night! He says it's work but I know he's really staying out as late as possible because it's so boring at home. As if he's the only one who feels that way!— Dear me—"

At that her mother would cast a wry grin and scowl at her. But before she could say anything, the little girl would exclaim, "You foolish girl! Come on, go to bed now. Little children who stay up late never grow, and they have to stay little for ever and ever."

The mother, utterly amazed at her daughter reading her mind time after time, would give in, saying, "This child is very bright, but she really tires me out!"

When she was a little older and her mother bought her a new toy, she would say, "This will keep her quiet for a while." Her mother, no doubt a little irritated, looked at her daughter, who would then say, "Why in the world does this child read other people's minds all the time. She's like a mountain witch. I wonder if people will come to dislike her like a mountain witch."

These are, of course, the kinds of things that her mother thought of often, and the child was merely verbalizing her mother's thoughts.

When she started going to school, the mother was, to a certain extent, relieved that she had times of separation from her daughter. But when she began to notice that her daughter ceased to read people's minds and became quieter each day, she asked, "How come you are so quiet now that you go to school?"

Her daughter replied, "When I say whatever is on my mind, people give me unpleasant looks, so I decided not to speak out any more. Grown-ups are happy when children act stupidly—as though they don't know anything. So from now on I've decided to keep grown-ups happy."

The mother responded firmly in a manner befitting one who had borne a mountain witch. "You say whatever is on your mind. You don't have to pretend. You're a child, remember?"

But the child merely regarded her mother with a disdainful smile.

All in all, the child performed well at school. On the occasions when she did not do well in a test, she would tear it up without showing it to her mother. Her mother would complain when she did not finish the lunch she brought to school, so on days when she did not have much of an appetite, she threw the remaining contents of her lunchbox into a trashcan on her way home. But so that her mother would not become suspicious, she left a little portion of it every now and then and showed it to her mother, saying, "The teacher talked longer today, so I didn't have enough time to finish it."

Time passed, and the child bloomed into maidenhood, but because her family was not well-to-do her mother could not afford to buy her expensive dresses. When the two went shopping together, the girl would purposely pick the dress her mother thought most adequate and pretended that she really liked it.

She would say instead of her mother, "I think this is really sweet. If I wore something extravagant at my age, I would give people the impression that I'm someone like the mistress of a rich old man."

On such occasions her mother would look at her with a slightly sad expression on her face. And on the way home, she would buy her daughter something way beyond her means. The girl would pretend not to notice her mother's impulse and showed a happy face to her as though she was genuinely pleased by her new acquisition.

The girl would assume whatever behavior was expected of her as though it was what came naturally to her, not only toward her family, but toward anyone by whom she wanted to be liked. When they wanted her to laugh, she read their minds and laughed. When they wanted her not to say anything, she remained silent. When talkativeness was desired, she chatted merrily. Toward a person who considered himself intelligent, she would act a little stupid—she did not overdo this, for usually this type of person thought it a waste of time to deal with stupid people—and as for those who were stupid, she appreciated their simplicity.

Perhaps because she demanded too much of herself and because she wanted too many people to like her, she had to spend an incredible amount of mental energy every day. So that before she realized it she had become antisocial, reading books in her room all day, avoiding being with others.

When her mother asked, "Why don't you go out with your friends?" she would answer with few words, "Because I get tired—"

The mother, too, began to feel fatigued when she was with her daughter. When she was not around her, she felt relieved. She began to long for the day when her daughter would find an adequate young man and leave her. In other words, the mother and daughter came to the natural phase of life when they would part from each other.

The daughter, too, knew that she was a burden to her mother—in fact, she had sensed that she was a burden to her as far back as she could remember—and she wanted to free her mother, as well as herself. At the same time, somewhere in her heart she held a grudge against her mother, a grudge which was sometimes so strong that she would feel surges of inexplicable rage. That is to say, she was going through the short, rebellious phase of puberty, but when she realized that her hatred and anger were directed at the cunning ways of her mother who had become her competitor of the same sex—that is, at her dishonest ways like taking advantage of her authority as a mother and avoiding direct competition—she became acutely aware that her mother had aged and that she herself had matured.

As a mature girl, she naturally came to know a man. He was an ordinary, run-of-the-mill type of man. Typical for one who had been doted on by his mother, he firmly believed that because his mother was of the opposite sex, he was allowed beyond all reason to express himself as freely as he pleased. When one such as he matures physically, the woman he marries has to be a substitute for his mother. For him, she has to be as magnanimous as a mother, as dignified as a goddess. She has to love him limitlessly and blindly like an idiot, yet at the same time have a spirit capable of being possessed by evil, like that of some sinister beast. Fortunately, however, he at least had the male characteristic of liking women.

Since the woman was gratified by the man, she came to think that she would not mind making all kinds of efforts to keep him happy. But this turned out to be very hard labor, for after all, every corner of his mind was transparent to her. If only one could not see another's heart, one would not become weary and would be able to live happily.

First of all, the man wanted the woman to be constantly jealous, so that she had to make every effort to appear that way. When another woman's shadow approached the man's life, she would act as though her presence made her competitive, and the man would be satisfied.

"Please don't go away from me. I can't live without you, you know that. I can't do anything by myself and I'm helpless when you're gone," she would cry as she sobbed and clung to him. And as she said the words, she would have the illusion that she really was a weak and incompetent creature.

Also, the man desired the woman to think of other men as something less than what they were, so that she had to close her eyes to the merits of other men and observe only their vices. But since the man was not excessively stupid, he did not allow her to denigrate others with idle speculation either. To please him, she had to make the right judgments, as well as be aware of all their vices, and indicate that even though they might have certain merits, these merits were certainly not to her liking. Thus every little opinion she expressed had to be well thought out.

On top of that, the man had the strange tendency of feeling pleasure in possessing all to himself a woman who was constantly being pursued by other men. Thus he tended to encourage rather than endure her affected flirtations. Perhaps deep down, all men long to become a part of the species of men we term "pimps."

To provide all the examples of this kind would take forever. In any case, at times the woman would forget to be jealous, or to flirt with other men. Or occasionally she was careless enough to express her true feelings about attractive men. At such times the man would become bored and think the woman lazy, thick-skinned, and lacking in sensitivity. Moreover, even when the woman succeeded in behaving perfectly to his liking, he would assert with the dignified tone of a sage who knew everything, "Women are utterly unmanageable creatures, so full of jealousy, capable of only shallow ideas and small lies. They are really just timid and stupid. In English, the word man refers also to human beings, but I guess women are only capable of being human by adhering to men."

Thanks to this irrational declaration of inequality, the two managed to live somewhat happily. Both the man and the woman grew old, and soon enough the man reached the age at which he would grumble all year long about something being wrong in this part of his body or that. He demanded that the woman worry about him all of the time and said that if anything happened to him he would be so concerned about her who would be left behind, that he would not die in peace. As she acted nervous and uneasy about him, she really became nervous and uneasy, until eventually she came to feel that he really was critically ill. For after all, unless she believed it he would not be at peace, and unless he were at peace, she could not feel that way either. Thus even though she hated nursing to such a degree that she thought she would die if she had to commit herself to it, she became a nurse just as a woman driven into a corner might sell her chastity. Observing the woman who now took up nursing, the man commended her, saying that nursing was an occupation truly in keeping with her instincts, and that as far as nursing was concerned, women were blessed with God-given talents against which no man could compete.

Around that time, the woman became exceedingly fat, so much so that when she walked just a little her shoulders would heave with every breath just like those of a pregnant woman. The main reason for this was that she was the possessor of exceptionally healthy digestive organs and consequently was constantly plagued by enormous appetites. But on top of that, she had the pitiful characteristic of wanting to make others feel good; even if she did not like it, she would eat up whatever was offered to her in order not to disappoint the person. Since other people thought that she just loved to eat, they would be terribly offended should she refuse the food that they offered her. On the other hand, her husband often boasted that he was a man of iron will. As she ate, saying, "Oh dear, here I go again—" he would cast a ridiculing glance at her; "You're such a weak-willed woman—" Even if someone put her heart into cooking something to please him, he would adamantly refuse if it was something that was not good for his health. In other words, his nerves were tenacious enough not to register shame at ignoring somebody else's feelings.

Because his use of words such as strength of will, insensitivity, and laziness so differed from hers, she would at times be overwhelmed by a sense of acute loneliness. She would come to fear not only her husband but many of the others around her as well, feeling as though she were surrounded by foreigners who did not speak the same language. Sometimes she thought she would rather live as a hermit in the depths of the mountains, just as she locked herself up in her room all day without playing when she was a little girl.

Far off in the midst of the mountains there would be nobody to trouble her, and she would be free to think as she pleased. The thought of extorting all those who tormented her in the human world made her heart beat with excitement: all those dull-headed, slow-witted people who could walk around with the looks of smug, happy heroes just because they were not capable of reading other people's hearts. If only she could say out loud like the legendary witches, "You just thought—didn't you!?" how relieved she would feel! It would be the sensation of slitting the skin around the temples in order to let horns grow, horns which are itching to grow out but cannot.

When she imagined herself living alone in the mountains, she likened herself to a beautiful fairy, sprawled in the fields, naked under the benevolent sun, surrounded by trees and grasses and animals. But once a familiar human being appeared from the settlement, her face would change into that of an ogress. He would stare at her, mouth open like an idiot, and utter coarse, incoherent, conceited words, making her fly into a rage.

On such occasions, her husband would appear, dressed shabbily like a beggar. He would wander about the abode of the woman who had now changed her appearance, and like a mischievous boy who had lost a fight he would mumble, "Without her to camouflage my unreasonable desires for me, I'd be done for—"

Listening to his voice, she would look at her face reflected in a clear spring. Then she would see that half her face was smiling like an affectionate mother, while the other half was seething with demonic rage. Blood would trickle down from half her mouth while it devoured and ripped the man's flesh apart. The other half of her lips was caressing the man who curled up his body in the shadow of one of her breasts, sucking it like a baby.

Now, as she became fatter, she began to develop arteriosclerosis, for her veins were put under increasing pressure. She found numbness in various joints of her body and suffered from headaches and the sound of ringing in her ears. When she saw a physician, he diagnosed that she was merely going through menopause. She was told in her early forties that she was suffering from menopause, and since then for over twenty years she had continued to receive the same explanation.

The man asserted that women were, as a rule, more durably constructed than men, their bodies and souls being more sturdily built. He pointed to a statistic that showed how women outlive men, and insisted that between the two of them, he would be the first to go. The woman thought that perhaps the reason women live longer has something to do with the fact that men end their lives of their own accord at youth, owing to war and other violent behavior, but since it was bothersome for her to prove this statistically, she just did not bring it up.

"That's right. Even though men are larger in build, they are actually sensitive at heart and more frail. That's why all women like men." As she said this, she told herself that the world would be a place of darkness without men, even though what she said was altogether fictitious, and continued stroking the man who complained that it hurt him here or there. In order to cook and feed him food as delicate as a little bird's, she spent hours day after day.

She knew that her own fat body did not have long to last with hardened arteries, but she could think of no other way to live than to provide food for the little bird of a man who believed that he was frail.

One morning, she examined herself thoroughly in the mirror. Her face was covered with wrinkles, giving her the appearance of a mountain witch. Her yellowed teeth were uneven and ugly like those of an aged cat. White frost had fallen on her hair and she felt chilling pain as though frost columns were noisily springing up all over her body.

She felt a slight numbness as though her body belonged to someone else. It was a stiffness related to the vague memory of her mother, long gone, far away. Somewhere, her flowing blood ebbed, and she felt dizzy. Suddenly a slight drowsiness attacked her, and when she came to herself her limbs were paralyzed and her consciousness dimmed as she felt various parts of her body gradually grow colder.

Customarily, she would have been up a long time ago preparing his meal. But finding her instead next to himself (they had slept alongside each other for forty years) face down and as stiff as a dead person, he became alarmed, and immediately straightening his body about which he had been complaining so much, he carried his wife to the hospital. Surprisingly enough, the physician who up until the day before had written her off as a case of menopause, now declared as if he were another man that she had the symptoms of cerebral thrombosis, and that if luck was against her she would only survive the next day or two. The man became totally confused, but he managed to pull himself together and decided that the first thing he should do was to send for their son and daughter, both of whom lived far away. The two children came immediately and with their father crouched around their mother who had now lost her speech.

Probably the next two days were the best two days of her life. The three of them took turns rubbing her arms and legs, and they would not leave it to the nurse to take care of even her most basic needs.

Even after two days, however, there was no drastic deterioration in their mother's condition; nor did it take a turn for the better. Her consciousness, however, became even dimmer, and she could no longer recognize the people around her. The uncertain physician said, "Considering her weight, her heart is strong. She may be able to hold on longer than expected." Soon the son claimed that he could not continue to stay away from work and that since it looked as though there would be no changes in the immediate future, he would return home for a while. With a gloomy look on her face, the daughter began to worry about her husband and children.

The poor man became anxious that he would not know what to do if his daughter left, so he pleaded with her to stay on. He sounded so helpless that the daughter, as worried as she was about her own family, reluctantly agreed to remain.

The daughter remembered the time when she had been critically ill as a child. Then her mother had stayed up for days watching over her. She thought that if not for this woman who lay in front of her unconscious, straying between life and death, she would not have been alive today. And this might be the last time she would be able to see her. Thinking through these matters, she hung on beside her. But when another two days passed, she began to wonder how long her mother would remain in her present condition, unable to converse and barely breathing, like a living corpse. She even thought that although the sixty-two years her mother had lived might be shorter than average, sooner or later everyone has to die, and that perhaps even if her mother went as she was from her present state, she would be considered fortunate that she could go watched over by her husband and daughter.

The daughter felt strangely uneasy when she remembered the story of the patient who survived for two years on intravenous feeding. She became worried whether her father's savings would be sufficient to pay the medical expenses should her mother survive as long. Even aside from the expenses, moreover, she thought that neither she nor her brother could afford to take care of her mother for such a long period of time, for they had their own families to consider.

She happened to think of her five-year-old daughter whom she had left behind with her mother-in-law. She remembered that at that age she herself had fallen ill and run a high fever for days, nearly contracting meningitis. Vividly, she envisioned her mother becoming frantic with worry and sitting by her bedside, cowering over her in their house which had become pitifully unkempt. Odd as it may seem, the impact of this memory led her thoughts away from her dying mother who lay moaning between life and death in front of her eyes, and made her concerned with the possibility of her own daughter falling ill while she was away. Unlikely as it may seem, she became plagued with fear at the thought of it.

Unaware of her daughter's worries, the mother survived another two days, occasionally staring into space with empty eyes and moaning something incomprehensible. The daughter woke up the third morning, too weary to climb out of bed after a week of intensive nursing. It was a dull, gloomy morning, typical of a cloudy day in the cherry season. She looked vacantly at the profile of her unconscious mother, who was also breathing quietly and who, with hollower cheeks, looked younger and beautiful.

When the morning round was over, the daughter, remembering that her mother's body was dirty, asked the physician if she could wash the patient. He instructed the nurse to do it and left the room. Soon the nurse came back, and in a very businesslike manner, carried out her duties as instructed, turning the patient over as though she were a log.

Timidly, the daughter helped her. Just when the patient was rolled over, stripped of her nightclothes soiled with perspiration and excrement, her eyes suddenly opened wide, staring at her daughter who happened to be standing right in front of her, holding her. She smiled faintly at her as light returned to her eyes. The radiance was like that of a firecracker, bright yet sad and ephemeral. Soon the firecracker died. The invalid lost the light in her eyes, and the saliva which had gathered trickled down the side of her mouth. Her throat went into a momentary spasm. The pupils of her eyes stopped moving, and then she was still. It all happened in a single moment.

At this sudden change, the nurse hurried off to call the physician. He rushed in and started to perform artificial respiration. He also injected cardiac medication through a thick needle into her heart. It looked more like shaking an animal that had failed in the middle of an experiment than dealing with a living human being. But in any case, it is certain that the people around her made various efforts to revive the pulse to her heart which had stopped.

The woman died.

No, it would be more truthful to say that she summoned up the last of her strength to suffocate her own self and body by washing down the accumulated saliva into her windpipe.

In the last smile she exchanged with her daughter, she clearly read her daughter's mind. Her daughter's eyes said to her that she did not want to be tied down by her any longer. "Mother, I don't need you to protect me any more. You've outlived your usefulness. If you have to be dependent on me, if you can't take care of yourself without being a burden to others, please, mother, please disappear quietly. Please don't torment me any longer. I, too, am preparing myself so that I won't trouble my daughter as I am being troubled by you. I'm willing to go easily. That's right. I ought to go easily. I never want to be the kind of parent who, just because she doesn't have the courage to come to terms with that resolution, continues to press her unwanted kindnesses upon her offspring." It seemed that her daughter, the product of her and her husband, possessed a strength of will that was twofold. Either she would overcome all temptation, exercise moderation, and live sturdily until the moment of her death at a hundred, or live haughtily and selfishly to the end, retaining the energy to kill herself at eighty. In either case, the woman was satisfied with the daughter she had borne and raised.

Through her daughter's face, she saw the son who was not there, walking among the crowds of the metropolis. He was talking to her with a crooked smile on his face. "Mother, I have incessantly chirping chicks at home. I myself don't know why I have to keep on putting food in their mouths. But when I catch myself, I'm always flying toward my nest, carrying food in my beak. Before I even think about it, I'm doing it. If I were to stop carrying food to them and stay close by you all the time, the human race would have perished a long time ago. In other words, for me to do as I do for them is the only way in which I can prolong and keep the blood you gave me—"

Next she looked at her husband, who was standing around absent-mindedly. This deranged old man, his head drooping, was touched by the beauty of his wife's naked body and absorbed self-righteously in the faithfulness that let him attend to his wife until the very end. The greatest happiness for a human being is to make another happy. She was satisfied with this man who had the capability to turn any situation into happiness, and she blessed the start of the second chapter of his life. At the same time, she thought she heard the pealing of her funeral bells.

With her own hands, she arranged her white shroud, left side under the right. In a dry riverbed, when she happened to look behind her, she saw somebody running away with his hair disheveled in the rushing wind. When she asked another deceased traveler whom she had not noticed before, the traveler answered, "He's being chased by a mountain witch."

Under the shroud which she had arranged, she felt the heartbeat of a mountain witch reviving, and she smiled. The heart of the mountain witch was throbbing as sturdily as ever. Only the blood vessels to transmit her vitality were closed, tightly, harshly, never to open again.

The time had come for the spirit of the mountain witch to return to the quiet mountains. The day had at last arrived when she would stand on a mountain ledge, her white hair swaying in the raging wind, sounding her eternal roar into the mountains. The transient dream of living in the human settlement disguised as an animal was now over.

The days she spent dreaming of living alone in the mountains, the sorrow she felt as a little girl when she first began to dislike humans, all came back to her and she shook her head. Had she lived up in the mountains, she would have been the mountain witch who devours humans from the settlement.

She wondered which would be the happier, to live in the mountains and become a man-eating witch, or to have the heart of a mountain witch and live in the settlement. But now she knew that either way it would not have made much difference. If she had lived in the mountains, she would have been called a mountain witch. Living in the settlement she could have been thought of as a fox incarnate or an ordinary woman with a sturdy mind and body who lived out her natural life. That was the only difference, and either way it would have been all the same.

Just before she took her last breath, it crossed her mind that her own mother must have been a genuine mountain witch as well. Strangely enough, when she died she had a mysteriously naïve face with the innocent smile of a newborn baby. Sobbing and clinging to this woman who died in peace, the daughter, with swollen eyes which told of her indescribable relief, said, "Such a beautiful death mask— Mother, you really must have been a happy woman." Her husband cried silently with wide open eyes full of tears like a fish.

十三

Yellow Sand

Hayashi Kyōko

As I opened the raindoors, the sky was a brown haze. The morning news explained this as yellow sand.

The dust that rose in the air in the interior of the Chinese continent, having crossed the Yellow Sea and the Korean peninsula on the westerly wind, seems to be blowing over a wide area from Kyushu to the Kantō and Hokuriku districts. The yellow sand borne by the forty kilometer per hour west wind takes one to two days before arriving in Japan. During the long two-day trip, heavier grains drop on the mountains of the Korean peninsula or into the Yellow Sea, and only comparatively light grains reach the Japanese sky.

Even so, almost all the sky above the entire land of Japan, it is reported, is shrouded in smoke, covered by a thick layer of sandy dust, as though wearing a cotton hat. On the day the yellow sand falls, however, there is an odd brightness. The grass field in front of my house and the trees of the surrounding hills are both bright, vaguely hazy like the light we see through frosted glass.

China is where I spent my girlhood. It was Shanghai, China, then called *Shina*. Part of the land which extends from Shanghai rises in the air in a huge cloud of smoke and descends in the air of the interior of Japan. I wanted the feeling of the dust directly on my skin and went out to the field. Stretching my arms wide, I inhaled deeply. Facing the sky, I shook my head violently a few times like a dog after bathing in water. The moist wind moved across my cheeks. There was the smell of the dirt of the continent reeking in the sun, though perhaps it was just an illusion.

Kōsa (1977). Translated by Kyoko Iriye Selden with the permission of the author.

In March, the yellow sand often blew in Shanghai, too. Just like the saying "Yellow dust of ten thousand yards," the yellow dust which starts around Taiyuan near the Gobi Desert sometimes turns the Shanghai sky the color of the earth. Even the Huangpu turns muddy and fuller than usual, and its level starts to rise. In the fields of the suburbs, rape flowers bloom. The sky, the river, and the earth merge in one yellow, and Shanghai enters spring with no distinction between heaven and earth.

Though dust is in the air, the wind is moist and soft to the skin.

In the field, too, the moist warm wind characteristic of the yellow sand days was blowing, just as I expected. In the misty wind that blurred the world, it seemed as though an event which should be called the original picture of my thoughts remained in its pure shape without any finger grease, as it was first perceived in my childhood days. Thinking of going where a richer yellow sand wind blew, I started to walk. The hills that surrounded me on three sides came to an end, and I reached a gentle slope.

Below the slope, I saw the rows of houses in town. Since the town formed a valley, yellow sand, more deeply hued than in the fields and hills, sank smokily to the eaves of the houses. I stopped in the middle of the road. Knotted currents of wind blew from beneath the slope. I closed my eyes and listened to its sound. The wind, slightly trembling as it passed my ears, brought back to me the hush of the rape flower field that I entered with Okiyo-san when I was young.

I first met Okiyo-san in the spring of 1937, the year I entered primary school in Shanghai. She might have been twenty-three or -four. On July 7 of that year, Chinese and Japanese armies clashed near the Marco Polo Bridge, igniting the Sino-Japanese War.

Both Okiyo-san's house and mine were located at the entrance of Hongkou, near the Garden Bridge. Hongkou was in a Shanghai district then under Japanese control. With the Garden Bridge over the Huangpu as the boundary, the international settlement lay across the bridge. It was controlled by England, the United States, and France. This side of the bridge was Hongkou, the Japanese quarter. Okiyo-san's house was a few minutes' walk from mine. In that area, red brick houses of the same style stood in rows. Perhaps because it faced the international settlement across the river, the appearance of the town reflected its influence. The houses were English-style three-story buildings with brown slate roofs. A big rectangular chimney stuck out of each roof, connected to the living room fireplace below. The roof had a bulge which resembled a horse's saddle. Though narrow, there was also an unfenced lawn that one could enter freely from the pavement.

Okiyo-san lived in this house with several prostitutes. Her fellow prostitutes were White Russians, she being the only Japanese. On the eve of the Sino-Japanese war, Shanghai brimmed with Japanese military men, and accordingly, there were brothels all over town. Japanese prostitutes were rare, however. Japanese women entertainers worked at Japanese-owned restaurants under supervision. They were divided into army and navy categories. They were not of course called prostitutes, and I don't know whether monetary transactions took place. Anyway, nobody ever heard of a Japanese woman like Okiyo-san who openly became a prostitute and engaged exclusively in the trade.

In the front of the yard was a street lined with plane trees. Beyond the pavement, the Huangpu flowed in a big curve around the Garden Bridge. Facing the river was a wooden bench for resting. Just across from Okiyo-san's house, too, was a bench. I was fond of sitting on this bench and watching steamers come and go on the river.

That day, after lunch, I sat on the bench watching the boats. People had gathered in front of Okiyo-san's house, which was usually quiet during the day. As I turned back attracted by the noise, I saw that they were coolies. They had probably missed out on longshoreman jobs at the port. In four or five minutes, they formed a human fence of nearly twenty people in front of the house. They were all shouting, looking at her yard.

In a corner of the yard away from the people was an acacia tree. It was the only tree in her yard. A man was leaning against the trunk of that tree. He was the man with rather dark skin seen around this area from time to time in the evening. He was twenty-three or -four, about the same age as Okiyo-san. He wore silken Chinese clothes. The coolies pointed one or two fingers at the man, shouting, "one hundred," "two hundred." The man, his arms crossed, shook his head in a lordly way, grinning. He did not respond. Finding the man adamant, the coolies raised their voices, the numbers increasing little by little by tens and fifteens.

The men seemed to be betting. In Shanghai this was hardly unusual. Whenever people gathered in a circle, there was always a bet going on. Anything would do for a bet. Sometimes it was an ant crawling on the street. Or some crickets in an unglazed pot, whose bottoms they tickled with the tip of a blade of pampas grass to see which could be made to chirp better. Everything that caught their eye became the object of a bet.

It seemed that here too, in front of Okiyo-san's house, betting had started. Interested, I crossed the tree-lined street and, mingling in the crowd, looked for the object of the bet. In the center of the grass was a bamboo couch. The betting seemed to center around that couch, but nothing like a betting object was seen on or near it. All I saw was the spring sun covered with yellow dust shining on the amber skin of the bamboo. A coolie loudly shouted, "Three hundred!" As if on cue, a few others, raising three fingers, shouted, "Three hundred" toward the third-floor window. Then they called "Okiyo!"

One window faced the Huangpu from the third floor. In our house, the third floor was the children's play area. If we jumped up with arms stretched in the garret, we could easily tap the beam, because the roof was low. The room was the size of a ten-mat room, but there was only one window. It was gloomy even during the day. On a fair day, however, it was pleasant: it felt like being inside a goldfish bowl, as the swaying reflection of the river waves made rings on the walls and the roof beam.

Okiyo-san's room was probably the third-floor garret. A blue curtain hung over the closed window. The coolies kept calling, trying to entice her to come down from that window. The man in silk raised both hands to quiet the coolies. The entrance doorknob turned slowly from inside. The door with a stained glass decoration opened slightly, and Okiyo-san came out as lightly as a small silver carp swimming in and out of river weeds. The coolies became quiet.

Okiyo-san wore a Chinese dress. On ordinary days, she wore a red Hakata-woven undersash over a summer kimono. I saw her in a Chinese dress for the first time. Her hair, too, had a big bulge in front like a visor. She was naked beneath the Chinese dress with a thigh-length slit on the sides. She wore satin Chinese shoes on her bare feet. In comparison with her hairdo neatly finished with oil, her bare skin seemed disarmed. With her smooth, pliant arms hanging naturally, Okiyo-san leaned against the door. The man in silk slowly approached the sofa. Seeing him move, Okiyo-san, too, walked toward the couch. Then she sat on the edge of the couch, legs decorously together. The man sat, too. The moment he sat, he drew her toward him, embracing her waist. He held her with his left arm and pushed her down on the couch. Okiyo-san, pushed down, beat him on the sides with both hands. Five fingers flitting apart, she beat him with an expression which seemed neither jocular nor serious. He clasped one wrist and then the other. He put them together with his hands, then held them with his right hand. After depriving her of the freedom of her hands, he held her legs between his.

Okiyo-san became still. The man also became still. *Wei, wei,* the coolies shouted. While shouting, one of them beat the man on the back with all his strength. At this signal, the man sprang up from the couch. Okiyo-san also sprang up with the same agility.

Swiftly smoothing the wrinkles on her dress, Okiyo-san smiled to the man, curling her red-daisy-colored lips. The man bowed to her, joining his hands in front of his chest.

The coolies clapped frantically. They threw the money, bid from one hundred to two hundred and up, clapping repeatedly. Seeing the copper coins and bills scattered around his feet, the man laughed aloud. Okiyo-san, too, laughed very loudly, beating her belly with her fists.

The bidding semed to have been on their union: whether the man would attack successfully, or the woman would defend herself to the end. The match had ended quickly, with Okiyo-san the loser.

It was then that I first saw the union of man and woman. I only saw it, without understanding the meaning of the act. The impression was not so strong as to haunt me, but the union accomplished in the sunlight was as refreshing as watching a pair of coupling dragonflies flying over the ears of rice.

The coolies had gone. The man in silk clothes was also no longer there. On the pavement devoid of people, I stood watching Okiyo-san. She stood with her back toward me but apparently felt my eyes on her back, for she turned around. She noticed me. Looking surprised, she asked, "Are you a Japanese child?" I answered, "Yes." "Were you watching?" she asked. "Yes," I answered clearly. Okiyo-san remained silent for a while. Then she said, "I see."

We started to walk toward the bridge. Okiyo-san sat on the bench. I, too, sat down, but not close by. A steamer moved toward the mouth of the river. At high tide, the surface of the river was swelling, as the water rose in a dull sheen. Okiyo-san watched the steamer going against the tide with her eyes narrowed; then she suddenly asked, "Do you think you could reach home if you added bulwarks of hardwood to that boat, hard as oak?" It would be dangerous, though it wouldn't be impossible to go home that way. But why invite danger when five- or six-thousand-ton-class ferries like the Shanghai and the Nagasaki were plying back and forth between Shanghai and Nagasaki? Once aboard the ferry, she could return to Japan as she wished, even the next day. Why did she think of the near impossibility of secretly returning, purposely choosing a small boat with bulwarks of all the steamers on the Huangpu River?

By setting up an impossible situation, Okiyo-san seemed to be trying to cut off her nostalgia for her country. The reason she had to cut it off probably lay in her past. That past, which forced her to cross the sea to Shanghai as a youth of twenty-three or -four to live away from her female compatriots and among foreigners who fled their countries, was perhaps so harsh as to never again allow her to live in her homeland.

As the days went by, her union disappeared from my memory. However, because she had asked if I thought it was possible to go home, before long I started to think of my country in connection with the end of the flow of the Huangpu River. Yet I knew nothing concrete about how I related to my country or how people lived over there.

Someone seemed to have told my mother. She knew that I had watched the union of bodies. She asked, "Were you watching?" "Yes," I answered. She stared straight into my eyes and asked, "Were you watching till it was over?" "Yes," I answered again. "Shame on her," she said angrily. "Isn't she a Japanese?—exposing herself in front of people; she is a disgrace to our nation." Then she told me that that kind of woman should be forcibly sent home.

This was the feeling of the Japanese adults in town toward Okiyo-san. When living in a foreign country, one was apt to feel that each person represented his or her home country. Since at that time national prestige was important, the Japanese residents especially were strongly self-conscious. Women were even forbidden to go out without socks or stockings. It was thought that Japanese women's skin should not be exposed to foreigners' eyes. My mother, too, went out wearing white *tabi* as though rich. Beggars, robbers, and even poverty were a disgrace to the nation, and were considered grounds for repatriation. This severity was not limited to thoughts about foreigners; they observed one another with an even severer eye. Such words as "traitor" and "repatriation" often characterized their conversation. Even children, when they fought with friends, said, "You'll be repatriated." However, the children did not necessarily realize that they were shouldering the weight of their country.

As extreme expressions became commonplace among adults, the clannishness among the Japanese intensified. On the other hand, the public peace of Shanghai was increasingly threatened. A town association was formed for self-defense. A vigilance committee was also organized. When compatriots live in a foreign country, the greatest security is in flocking together.

Okiyo-san, though Japanese, was excluded from the group. She didn't seem to mind too much.

In April, half a month after I saw the union of bodies, I entered primary school. One day, on the way home from school, on the bridge over the creek, I met Okiyo-san. The bridge spanned the Hongkou Creek which separated my house from the school. In the middle of the bridge stood two huts painted green, facing each other. Marine sentinels stood guard there armed with bayonets. Behind one hut a long line of Chinese people had formed on one side of the bridge.

In the summer, cholera raged in Shanghai. All residents, regardless of nationality, had to be immunized. Japanese were supposed either to receive injections at hospitals especially designated for them or to call a doctor to the town association to immunize the entire town. My mother was among those who had received a shot from a doctor at the town association the day before.

Only Chinese people got shots on the bridge or on a street corner. Okiyo-san was standing in the all-Chinese line. In the same Chinese dress as on the day of the union of bodies, she moved one step forward each time the line advanced. In her Chinese dress, she seemed no different from the others. The people before and after her didn't seem to notice any difference. They spoke to Okiyo-san. Yet, although her skin and bone structure were the same as those of the Chinese people, somehow she was different. To me who knew her to be a Japanese, Okiyo-san alone stood out. Okiyo-san looked pitiable to me, shut out by her compatriots and joining the foreigners' line. Going near her, I softly called from behind, "Okiyo-san." She turned back, and as she noticed me, said, "Go home quick," with her hand on my school satchel.

"I'll wait for you," I said.

We waited nearly half an hour before it was finally her turn. The army surgeon who stood with his legs apart like a Deva king grabbed her left arm and abruptly stabbed it with a needle without even sterilizing it. Injecting the liquid in a flash of a second, he hurriedly pulled out the needle. The point of the needle seemed to have been worn, for a black drop of blood came out after it was pulled out. "Does it hurt?" I asked. "Huh?" The surgeon inclined his head, and asked Okiyo-san, "Are you a Japanese, too?" Okiyo-san didn't answer. Rubbing the drop of blood into her arm with her palm in the same way as the Chinese did, she started to walk in the opposite direction from home.

Across the bridge and past the primary school, we came to another bridge over the same creek. As we crossed this bridge, we left the Japanese town of Hongkou. There were few houses, few people, and we came to an open field. As we walked further, there were no more houses, only the field extending to the horizon. An asphalt road divided the field. It had been made by the Japanese army in order to carry materiel to the Chinese interior.

Okiyo-san walked in the middle of the wide road. Rape flowers bloomed all over the field, and here and there amidst the flowers white-walled houses could be seen. These small houses were not for people; they were not even as tall as children. They were only as large as one tatami. The roof of each hut was made of lusterless black slate. Right under the slope, one window was open, the size of a postcard. The white-walled houses were said to be Chinese people's graves. To what class of people these graves belonged, and whether or not this house-like style was limited to the Shanghai area, were unknown. The graves had neither the names of the dead nor the dates of their deaths. All that were there were white-walled graves. They were not formed into clusters like Japanese people's graves; one by one, they stood separately on the spacious earth.

"Come and see," Okiyo-san said from among the rape flowers. I went into the field of blooming rape. As told, I peered into the window of a grave. There were no wooden pieces of a rotten coffin, no bones, no clothes—nothing attached to the dead. Weeds growing high enough to brush the ceiling filled the grave, stretching, their heads together, toward the window opened for the dead. The grass was deeper green than the green of the rape plants bathing in the sun.

"All I see is grass," I said. "That's what man is," said Okiyo-san.

Pressing down the rape flowers with her hands, Okiyo-san sat on the flowers with her legs outstretched. Then she lay back on the flowers, her hands pillowing her head. "The sky's pure yellow," she said. I stood in the flowers, watching the rape field stretching endlessly. The yellow of the flowers spread from around my breast into the sky and blazed as it met the sky at the horizon.

In the yellow color fused without break, the scattered grave windows were black holes. The hollow where Okiyo-san lay also had opened a big black hole in the earth.

The season of the yellow sand being over, Shanghai in July was in summer. Since parting in the field of rape flowers, I hadn't seen Okiyo-san. That we two had lined up in the column for immunization, and that we were in the rape field, my mother knew.

A Japanese neighbor had seen us and alerted her. I had done nothing to be blamed for. But just because Okiyo-san was a prostitute, my mother said, "It's your affair if you become a delinquent." She asked me what we had talked about in the rape flowers. I didn't answer. Suppose I told her, "She said 'That's what man is,' pointing at the weeds in a grave," my mother wouldn't understand, I thought. Only in the light of the spring field where heaven and earth joined, the words made sense.

The activities of anti-Japanese elements increased day by day. In the bustling quarters of Hongkou, an army lieutenant was shot with a pistol in broad daylight from across the street. Since he was shot at close range, the event shook up the Japanese all the more.

M Products where my father worked always ordered its employees' families to leave before an area was exposed to the disasters of war, as was to be the case at the conclusion of the Pacific War. Thanks to their secret information, about the time the family finished evacuating, war inevitably broke out in the area.

Before the skirmish at the Marco Polo Bridge on July 7, employees' families had already been ordered to leave. We were prepared to return to Japan as soon as we could obtain boat tickets. My mother said to me who hadn't seen our country, "It's safe at home; there we won't have to flee as refugees."

Partly owing to the threatening atmosphere, I was forbidden to go out of our yard. That day, too, I was playing with my thumb on the mouth of the sprinkler in a corner of the yard, spraying the water to the street, when I saw Okiyo-san walking. I stopped my water game and went to her. With a red undersash over her summer kimono, she said in a low voice, eyeing the windows of my house, "Would you come and play later? I'll give you biscuits with sugar crystals." Then she asked, "Are you going home?" As I nodded, Okiyo-san, too, nodded a few times, saying, "I see."

It was two hours later, about four o'clock in the afternoon. Okiyo-san hanged herself. Slipping away from my mother, I went to her house on the tree-lined street. A bigger crowd of people than that on the day of the union had formed in front of her house.

A black Municipal Police car was parked along the walkway. This car was, so to speak, a cleaning car: it collected everything including corpses of human babies discarded on the street, sick people on the road, and dead cats and dogs. I went among the coolies so as not to be seen by Japanese neighbors, and looked up at Okiyo-san's window.

"It seems she was found hanging from the beam," I heard a Japanese woman say. "I hear she hanged herself with her undersash," I heard another female voice. "How frightening—undersashes are strong, of course," I heard my mother say from the crowd unexpectedly nearby. I ducked down in the ring of people.

Okiyo-san, it was said, was found dead hanging from the beam of her garret. Its ceiling was low. Okiyo-san was not large, but if she had hanged herself with an undersash dangling from the beam, her toes would have reached the floor. How did she hang herself from the low beam? Listening to the talk of my mother and others, I recalled Okiyo-san's voice when she said she would give me biscuits. Why Okiyo-san, who had made that promise just a while ago, suddenly committed suicide, I couldn't understand. At least when she made that promise, she must have meant to be alive two hours later. If I went to her room, there would be biscuits with sugar crystals ready for me.

I had to see the biscuits wrapped in tissue paper, placed on the table or stored away in the tea cabinet.

The knob of the door decorated with stained glass turned slowly, and a plump, red-haired prostitute opened the door.

The staircase leading upstairs was visible from the entrance. A man in white came downstairs holding a stretcher high. It seemed that his steps didn't harmonize with those of the man behind who held the rear end of the stretcher. The stretcher swayed right and left at each step on the staircase.

Okiyo-san lay with her head toward us. Her hair showed from behind the man in white. She came out on the stretcher to the crowded yard. Neither a white cloth nor a blanket covered her body. Her arms powerless on the stretcher and her head stretched, she was dead. Her body looked longer than when she was alive. On that long body, she wore her familiar summer kimono, her usual undersash around it.

The men in white put Okiyo-san's stretcher directly on the car floor. Then they noisily shut the folding doors. The car with a small iron-grilled window drove off in the still bright summer city, carrying Okiyo-san. The car presently became a black speck and disappeared.

Two days after she hanged herself, our family returned to Japan. There I heard the news of the outbreak of the Sino-Japanese war.

The yellow sand that reached Japan covered the sky all day long. I was restless. I went outside many times and walked to the slope. In the blowing wind, I felt peaceful. I stood on the slope a long while, looking at the sky and the town. At the bottom of the slope in the gathering dusk, several girls were playing.

The yellow sand grew deeper as time passed. In time the sky, the town, and the girls all sank into the yellow sand, dispersed as countless black specks. It resembled the rape flower landscape which I saw with Okiyo-san, but in the landscape at the foot of the slope, the brightness of that day was missing.

十四

In the Pot

Murata Kiyoko

I

It was evening, when it was still light outside.

As I sliced vegetables in the kitchen, Grandmother came in through the back door with the things she had picked from the garden.

Grandmother's garden was small, and so was everything produced there—strawberries, cherry tomatoes, black corn. . . . She would pick those dainty things, and at that just a little at a time.

"Take a look, Tami."

In her palm were four or five deep green, thin peppers.

"If you slice these thin, real thin, and scatter them when the pot is just about ready, it'll add color to the soup."

I took the green peppers from Grandmother's hand. They were warm, probably from the heat they had absorbed all day under the summer sun. I put them on the cutting board and started to slice them.

As I moved the kitchen knife, music from the tape deck in a distant room reached my ear. My cousin Minako had the bad habit of playing the stereo while studying—although she never played it when she was lying on the floor reading comics and magazines.

Near me a big black iron pot was on the fire. In it, vegetables were boiling, trembling in the many little bubbles that surrounded them. A white vapor with the aroma of chicken began to rise, and a look into the pot somehow reminded me of the hot spring I had visited long ago on the sixth-graders' field trip.

Nabe no naka (1987). Translated by Kyoko Iriye Selden. This story has benefited from the translation assistance of Hiroaki Sato and L.

The stereo wasn't the only sound that reached the kitchen. I could hear something else, too: the old pedal organ from the gloomy room behind Minako's. It was the organ Grandmother had used when she taught elementary school long ago. Damaged by moisture, a few keys had stopped playing. So "Wild Rose" came wafting through, missing notes here and there—a wild rose with holes.

At the organ was my cousin Tateo.

It was an uncommon name—Ta-te-o; it meant "vertical man." While listening to his organ, I dropped the sliced green peppers into the pot. Coinciding with the final section of "Wild Rose," played as it was on an organ missing some keys, the green peppers fell onto the pieces of chicken in the pot as if full of sadness.

My brother Shinjirō returned through the kitchen door. Every day he took his school books and ran away from the strange noises of the house to the yard of a nearby shrine, where he took a nap.

"Did you study?" I asked as I removed the pot from the fire.

"Yeah," he answered innocently.

He must have gotten a good sleep today. There were traces of grass on one cheek; that must have been the cheek he had slept on.

"I can study outdoors best," he said, disappearing into a corner room.

This was at Grandmother's house in the country.

The four of us, her grandchildren, came here in the last week of July when school was out for summer vacation. My little brother Shinjirō, myself, and our cousins Minako and Tateo.

Our grandmother was eighty. She was all skin and bones, but still very healthy.

Swaying in the green rice paddies that early afternoon, her small, white parasol immediately impressed us. Grandmother had waited a long time for us to come, standing on the path between the paddies. Wearing a skirt and shoes, she held the parasol that had been stored for years at the bottom of her chest of drawers.

Grandmother led us through the rice fields to her house. Once we entered, she scurried back and forth for a while between the kitchen and the dining room, now slicing cool watermelon, now offering cold wheat tea. Then she handed us an airmail letter.

It was a foreign letter to Grandmother from Hawaii. To begin with, this was the reason for our stay with her out in the country. Assuming an important air as if he were our representative, Tateo took the envelope with its bright blue and orange edges. The moment he glanced at it, Tateo must have felt as excited as Shinjirō, Minako, and I did. Contents aside, the blue and orange stripes alone thrilled us.

Opening the airletter, we saw a Japanese handwritten script that was even slightly clumsier than my brother Shinjirō's:

Greetings
Mrs. Hanayama Sanae

I am the son of your younger brother Haruno Suzujirō

My father came to Hawaii from Japan in 1920 with a big dream he lived with pineapples and he will give me a big farm and will die soon the runaway wishes to see you please come see my father I beg of you

Sincerely
Clark

Thanks to Grandmother's long life, Clark's letter had reached her as he had hoped.

Hanayama is Grandmother's married name, Sanae her given name.

She was not particularly pleased to find her younger brother alive, or that he was probably a millionaire on a pineapple farm. He belonged more than sixty years in the past.

Grandmother and Suzujirō might have been at odds, Tateo suggested. "Women are vengeful. They won't visit an enemy even if he's dead."

She was even less excited about discovering her brother's whereabouts, than she was about the outcome of this discovery: our parents, Minako's parents, and Tateo's parents had gone to Hawaii to see Suzujirō and Clark, leaving Grandmother to care for us in their absence.

Our parents had already exchanged many phone calls all through July, talking long and hard about this Hawaii thing; then they called Clark in Hawaii to establish their relationship as cousins. Our fathers told their employers that a relative was ill in order to take vacations from work, and probably also added the ensuing funeral and all kinds of other excuses, while our mothers asked neighbors and friends to keep an eye on our homes before they sent us off to Grandmother's.

Though indifferent to the fuss around her, Grandmother came to meet us in the rice field wearing a skirt and shoes.

"Because things are this way," she said, looking into our faces, "you should stay here through the summer. Now's the only chance for you all. When you grow up, you'll be going separate ways. Enjoy living together for now."

Since our arrival at her house, that was the only thing Grandmother ever said concerning her younger brother in Hawaii. But more important, she was now very busy.

In the late afternoon of the day we arrived, Grandmother started a major search for something, opening all the cabinet doors in her spacious kitchen. She placed large, old pots in a row on the wooden floor. It seemed that there was a pot that she could not find, no matter where she looked. She turned her neck and tilted her head, trying to recall the time when she had used it every day.

"One big pot . . . , one, two, three middle-size pots, and one, two woks . . . and . . ."

I sat on the floor next to those pitch black iron pots. The wooden floor was cool and damp as always.

"The steamer pot's here, but, well, where's its lid?"

She lifted every one of the piled-up pots and looked inside. She won't find the steamer lid, I thought. That's because it—a substitute lid for an old *miso* vat—was on top of the cupboard where she had looked a while ago. As long as I did not tell, that lid would not come down from there. I should not tell her, I thought, that her grandchildren, already sixteen or seventeen, no longer cared for steamed sweet potatoes and pumpkin cakes.

"Well, I have my work cut out for me from today on," said Grandmother, giving up her search. She stretched her back. "The pots are here. The grandchildren are here. It's time to begin."

Grandmother carried one of the old pots to the sink. The small pot she had been using until yesterday was stored away. Just then, the sound of the organ came from the inner room. Her hand still on the tap, Grandmother sighed. Following the old pot, the old organ had begun to play.

Tateo was fooling around, playing with only his right hand. He was nineteen, the oldest of us four. Having just passed into college this spring, he was absolutely determined not to study for the rest of the year. So he had been walking around the house just looking at things in a laid-back manner. The discovery of the organ seemed to bring a feeling of contentment to both Tateo and Grandmother.

"Try playing 'Wild Rose,' " Grandmother had asked.

But . . . but the food that she prepared to the sound of Tateo's organ music was absolutely dreadful.

It was a dish of pumpkin, Kōya-style tofu, and chicken, cooked together. The chicken and tofu were so black that they were indistinguishable in the pot. As for the pumpkin, it was no longer to be seen. Melted out of shape, it had served the unintended purpose of thickening the sauce.

Our tongues shriveled up from the saltiness of Grandmother's overuse of soy sauce. In total silence, we moved our mouths. Having carelessly thrown a big morsel into my mouth, I had nowhere else to put it.

With the awful-tasting food on it, my tongue seemed to wander around, lost.

Grandmother alone chewed busily.

"Delicious, delicious. It's so special when we eat together, isn't it?" she said, and quickly finished two small bowls of rice.

Already on the first day of our stay out in the country, we despaired of Grandmother's cooking. The second day was worse; the third even more serious. The cause of the poor quality of Grandmother's cooking, I thought, was most certainly her false teeth. Probably she had always cooked soft food for herself and eaten it alone. And since her false teeth covered as far as the roof of her upper jaw, she had no doubt lost the ability to sense even half the taste.

In the middle of supper on the third day, Shinjirō suddenly faced her.

"Grandmother," he said. "Let Tami do the cooking from tomorrow. I think that'll be better for everyone. You'll be less busy, too."

Minako, Tateo, and I lowered our heads and closed our eyes, holding onto our rice bowls. Grandmother looked sad, her eyes falling on the pitch-black fish, overcooked and shapeless, on her plate.

And that was why, while I felt so sorry for her, Grandmother gave the kitchen over to me.

Grandmother loved stewed dishes flavored with soy sauce.

The menu I prepared in her place the following day was shrimp cooked with lotus root. With Shinjirō as my porter, I went down to the town at the foot of the mountain to get the ingredients.

Holding the serving bowl containing my stewed shrimp in both her hands, Grandmother was pleased to tears.

"Look what Tami made," she said, lowering her head toward the bowl as if in prayer.

The next day I fixed eggplant stewed with beef and konnyaku. The day after that was fried leek and pork, and the day after that I prepared chicken liver with scallions and green peppers. The pot was emptied every day.

Tateo, who had passed the entrance exam hurdle in the spring, spent his time playing the organ, reading books, and usually took a nap. Minako, a high school junior, was unable to study seriously; she spent her time curling her hair, polishing her nails, or reading comics.

My younger brother Shinjirō, a junior high student, kept going out. At home, my parents always sent me to the road at dusk to meet him when he was out later than he was supposed to be, but this was a rural area with nothing but a temple, a pond, and bamboo groves and there was no need to worry that he might be picked up by delinquents.

. . . Yesterday I stewed carrots, lotus root, and taro. What combination should I prepare tomorrow? . . .

This was the highlands, cool even in the summer. The heat rash that had broken out on Shinjirō's back at the beginning of the summer completely disappeared in the first four or five days after our arrival.

The wind that blew through the kitchen was dry and cool. Minako watched me sympathetically as I struggled every day with Grandmother's big old pot.

"Coming all the way down here to be a cook," she would say. "It's your own fault, taking over for Grandmother so easily."

But since my mother used to have me cook in a hot kitchen the moment school was out for summer vacation every year, I didn't find the kitchen chores in this cool house as bothersome as Minako imagined. From childhood I was the type of girl who asked her mother to please let her slice potatoes.

Minako was my age, seventeen, but she wasn't as good at peeling potatoes. But then, I couldn't curl my hair as neatly as she could. In a sense, we were girls who could each do half of the important things.

So we said to Tateo, who rose every day only at meal times, and otherwise slept with a book on his belly in the living room, on the verandah, in the shade of a tree in the yard, or anywhere else he could find a spot:

"Both of us will be your brides. So get up, or you'll be a poor man if you just lie around like that."

"Marriage between cousins is dangerous," he said, laughing, looking at us. Then he added, "If it's all right that it's dangerous, I'll take care of you."

"Why?" I asked.

"How dumb can you be. Don't you even know about Mendel and his cross-fertilization of peas?"

Cooled by the soft breeze under the blooming paulownia tree, Tateo, still lying on his back, started to lecture us: purple pea flowers were dominant, the purple and white flowers separated in the second-generation hybrids, R being the gene of the purple variety and r being the gene of the white. . . .

Then he told us the frightening tale about one sick purple pea flower that mingled with Queen Victoria's blood, and finally planted purple poison in generations of princes and princesses.

Minako and I ran from Tateo's lecture and walked home from tree shade to tree shade in the yard.

"If you help too much with the kitchen work," Minako warned me, "you'll be invited back next year, and the year after, and forever. Be careful." Minako was seriously worried for me.

"Do you think Grandmother'll live that long?" I said in a low voice, looking around.

"Oh, yeah," Minako answered with great confidence, "she's definitely not going to die for quite a while."

True, compared with other old people, Grandmother wasn't yet getting forgetful, nor was she troubled by neuralgia—which is rare for an eighty year old.

"I think it must be the result of the crossing of extra strong peas," Minako said. I nodded in agreement.

When we entered the house, the humid, cool air was moving quietly. I felt as though I were in water. At the bottom of that murky water where there was little light, Grandmother was stringing the snow pods that I was going to fix for supper.

I planned to cook snow pods, Kōya-style tofu, and chicken. With a pile of snow pods from the garden spread out on the newspaper on her lap, Grandmother was carefully removing the strings, pod by pod. Between the tips of her wrinkled thumb and forefinger, a green string stretched like a thin thread.

"Tami?" said Grandmother, still looking down. "These snow pods look very tender."

I saw the crown of her head. Her hair was quite thin. Her thin hair sent me abruptly into a totally unrelated vision: I saw a bamboo curtain hanging by a window on a summer night, swaying slightly. There was a light on, and the inside of the room was clearly visible through the curtain. The top of Grandmother's head was sort of like that.

It wasn't that she had no hair but the hair she had no longer played the role of covering her scalp. More ashen-white than flesh-colored, the lifeless skin visible beneath her hair reminded me of thin plastic bags. Her fragile-looking hair, so worn that it looked about to break, was stroked over the scalp, in a clear pattern left by the teeth of her comb.

Her hair was such a lonely sight. Inside the bamboo curtain, in my imagination, was a tatami room lit with an electric lamp, but there was no sign of people or voices; even the air was hushed. The atmosphere suggested that the light would go off in the time that I blinked.

After that only the bamboo curtain of the empty house, now reduced to a mere image of horizontal black stripes, would be swaying in the night wind. This was the sad scene that appeared in my mind's eye.

"Okay, Tami, carry this to the sink please," Grandmother said, lifting her head. A mound of bright green stringless snow pods transferred from her hands to mine.

I vigorously washed the pods, and sliced them in half with the kitchen knife. Then, I threw them in Grandmother's big, old pot, and fried them with chicken at a high temperature. When hot oil spattered in the pot and the snow pods started to dance, my cousins came home—Tateo from beneath the flowering paulownia, Minako with the laundry from the backyard, and my brother Shinjirō from his nap in the tree shade in the temple.

One night, after each of us had laid out the bedding in our rooms, we gathered in the dining room to say good night. I suddenly realized that I had left my brother's pajamas still hanging in the backyard.

Darkness was deep in the yard at night, and the clothes rack was near a big ginko tree. I knew that at night the branches of that tree hid the sky, blocking off the moon and stars.

"If you bring the laundry after dark, it's going to be wet with dew," Grandmother said, bringing Grandfather's old yukata from her room for Shinjirō.

My brother shrank from the yukata at a glance, saying, "I'm sleeping in my underwear. I don't need it."

"Come on, don't cop out on this," Tateo said, delighted by the old yukata. Jumping at Shinjirō and holding him down, he urged Grandmother, "Quick, quick," and the two of them quickly forced the robe on him.

Shinjirō, tottering out from between Tateo's and Grandmother's arms, had turned into a funny boy we had never seen before. The faded cotton yukata reached only to his knees.

Hitting the tatami, Minako and I burst into laughter, while Tateo fell on his back, laughing uproariously. Shinjirō just stood there stiffly, muttering, "Damn you guys." But there was one person in the room who didn't laugh: Grandmother.

She looked up at Shinjirō as if there were nothing funny about how he looked. Then shaking her head, she said softly, "I can't believe how much you look like Jikurō!" She made him walk around a circle in front of her.

We had never heard of a boy by that name.

"Jikurō was my youngest brother."

That was how Grandmother started to tell us about this boy. Jikurō had gone insane before entering middle school, and was confined to a room with wooden bars, she said. She talked of him with far more passion than she ever showed for Suzujirō, who was perhaps now dying in Hawaii.

"Long ago, somehow there were cases of insane children all over the place, but Jikurō was a quiet, good boy," Grandmother said.

"A room with bars—so it's like a jail, isn't it?" Tateo asked.

"Yes. It was a proper room with a tatami floor, but without the bars he would have wandered away."

"Poor boy," Shinjirō muttered with a sigh.

I imagined a lonely boy about Shinjirō's age sitting all alone in a corner of a barred tatami room. Everything, including outdoor noises, human voices, and sunlight, must have floated past him like a pale film strip, at some distance.

We became strangely quiet, and looked at Shinjirō. Through his figure, we were gazing at this insane boy named Jikurō.

"Did he spend the whole day in that room?" Minako asked.

"He was such a quiet boy."

"What was he doing all day?" I asked.

"He was studying," Grandmother answered proudly, straightening her bent back.

"What was he studying?"

"He was writing all sorts of characters on paper."

"What kind of characters?" asked Shinjirō, who had seated himself in the center with his legs crossed.

"Chinese characters like *hand* or *foot.* He often wrote the character for *neck,* too."

We looked at one another in silence.

"He was always fond of studying. Even after he went insane, he had fine penmanship."

"What else did he write?" Shinjirō seemed interested.

Grandmother, searching her memory and counting off on her fingers, recited, "*Eye, ear, mouth, nose . . . ,* then . . . *head, neck, chest, belly, back, bottom, leg . . . , hair, nail, bone . . .*"

"Oooh," Shinjirō groaned. "What a weirdo!"

Jikurō must have been obsessed with the names of the body parts.

"Didn't he write names of flowers and animals?"

"No." Grandmother shook her head.

"Did he write with brush and ink?" Tateo asked.

"It was a long time ago, so he wrote with a brush on paper. He'd sit in a dark space in a far corner of the room with bars, leaning on a pillar, and raising both knees, place the paper on them. He'd write just one large character, say, *neck,* on a sheet of paper. He'd spread the finished sheets all over the tatami mats. Paper was precious, but my father always gave Jikurō plenty."

Both Tateo and Minako fell silent. Shinjirō scratched his neck.

What a strange relative we had. I recalled Mendel's purple and white pea flowers that Tateo had talked about. I thought how lucky we were that Jikurō's purple flower hadn't come around to us.

As our conversation continued, the little bits of darkness that lurked in every shadow in the spacious house began gradually to occupy the center of my mind. I thought of the shadows of the chest of drawers, the cupboard, and even a tiny tea cup. Minako started to poke my foot.

So, making the excuse that we were sleepy, we said good night to Grandmother and Tateo, and fled from the story of the boy with the short kimono.

When we got out to the hall, which faced the yard, a large star like a pale blue eye was looking down at us from the sky. The mere sight was enough to shrink my heart.

"Hey," Shinjirō's voice whispered behind me.

I tried not to let his voice tempt me to look back at him.

"Wait." Shinjirō came after me.

"No!" I said walking faster. "Don't follow me!"

He sounded worried. "But we're next door to each other."

Catching up and walking with me despite my reluctance, he whispered in a low voice. "Sis," the boy in the yukata said insecurely, "let's sleep with the sliding doors open between our rooms tonight."

That night, with the sound of my brother's quiet breathing coming from the other side of the half-open sliding doors, I felt somewhat strange. Shinjirō was asleep as the brother I was used to, having taken off the yukata and put on an undershirt. I looked at his face through the opening of the screen doors.

I tried to superimpose on that face, like a mask, the imaginary face of the crazy boy. Then, those lips of Shinjirō, which he pouted whenever he got angry, I felt, would be closed quietly so that no obnoxious words would ever be uttered again. Those bright eyes, which sometimes glared askance at me, would always stay in a wide open stare.

He would be like a quiet doll. If he turned into that kind of little brother, I would be able to be a gentle big sister to him, too. This was my thought as I watched my brother's sleeping face.

I would write lots of Chinese characters for him: *eye, ear, mouth, nose . . . , eyebrow, eyelid, cheek, lip, chin. . . .*

Yet . . . yet, before that, shouldn't I practice Chinese characters a little bit more?

At that point, I finally came back to reality. The report card I had brought home from school before summer vacation flashed through my mind.

II

A letter came from our parents in Hawaii.

According to the letter, Suzujirō was overjoyed by the Japanese visitors and had started to recover remarkably. Furthermore, when they showed him the photograph of Grandmother that they had brought with them, he was so moved that he sat up in bed, got out his glasses, and gazed at it.

Thus, the plans of my father and my uncles to file the death report of a relative as a legitimate excuse for their prolonged absence from work were spoiled. On the brighter side, Suzujirō, who owned a large pineapple farm, was pleased to pick up all the expenses of their visit to Hawaii. His son Clark had also sent a letter, and it read as follows:

Hello
Our Japanese aunt

Thank you for sending many of your children my father is much better he is talking about you every day he is asking them about you we will keep them a little longer for my father's sake stay well respectfully

from Clark

Grandmother listened, with her head downcast, as Tateo read the letter. When he finished, he shook his head.

"Beats me," he said. "Grandmother's younger brother's a millionaire. I'm definitely going to Hawaii next year. I'll just hang out for a month or so."

Shinjirō looked away. "I don't want to go to such a place," he said. "Hawaii's a place where nerds go. Only nerds and honeymooners go to Hawaii."

"All right, you said it. Don't forget those words. I'll gladly be a nerd—I'll go to Hawaii, and I'll make friends with Clark's children before I come home. I'm sure Clark has children like us. I'll make friends with those gold- or red-haired cousins, then ask them to introduce me to their boyfriends and girlfriends."

At Tateo's vision, Shinjirō's feelings instantly changed.

"You think there are such people?" he asked seriously.

"It'd be unnatural if there weren't. Generally our family's prolific."

"Do you think I can have a conversation with those people with my kind of English?"

"That depends on your effort. As for me, there's nothing to worry about."

"I'm going to start studying English tomorrow."

Shinjirō ended up readily agreeing with Tateo. But it was not just Shinjirō who was enthusiastic about Tateo's story. At the first mention of blond boyfriends, Minako had already put her hands on her chest and shouted, "Me too, it's going to be Hawaii next year!"

"How about you, Tami?" Minako asked, as if she were talking about going the next day.

"Of course, I'll go too."

"Right, we'll all visit Hawaii next year," Minako said in an excited voice.

After that, our discussion of Hawaii became quite lively.

"Do you know who discovered Hawaii?" Tateo said, looking at Shinjirō.

"Wasn't it Captain Cook?"

"What's the capital?"

"Cut it out. I'm in junior high. Honolulu, of course."

"The aboriginal tribe?"

"The Kanakas. Don't you know I'm good at geography?"

Grandmother remained disinterested in the boys' banter. It was as if she were determined that Hawaii should have nothing to do with this house.

But how strange, I thought. There was Jikurō in the barred tatami room. Then, on the other side, there were blue-eyed children on a pineapple farm—or there probably were. Jikurō and the Hawaiian children were separated by time and geography, but in fact the distance wasn't all that great. It was just as Jikurō had seemed right beside us when Shinjirō whispered to me that night, "Let's sleep in the same room, Sis, I'm scared."

I thought about the wind that blew through me when I was walking through the sheaves of rice in the field. The rough wind stung my neck and arms and made me itch. I wondered if the wind that blew across the pineapple farm could sting like that.

Despite our excitement, Grandmother remained silent, her head cast down, looking oddly uncomfortable. Then, she left us without any of us realizing it.

We noticed her absence in the middle of our chat. Tateo looked around and said quietly, "I wonder if Grandmother doesn't want to go to Hawaii."

"Maybe she feels lonesome about being left alone after we all go to Hawaii next summer."

"Let's just take her, whether she wants to go or not," said Shinjirō, who had first refused to be a nerd, swinging his flexed arm. "I'll carry her. Easy."

"She probably only weighs thirty kilos," Minako laughed.

"She can't weigh that much," said Tateo. "Twenty-five or -six, probably."

How could a human being weigh so little and continue to live? Minako and I looked at each other while Shinjirō collapsed backwards in amazement.

Grandmother did not return until evening. Minako and I worried that she had run away from home, but Tateo, saying it was all right, stretched out on the breezy verandah. But he was restless with worry, his raised leg, crossed high over the other, bounced nervously up and down. Minako, Shinjirō, and I went without Tateo to search places that Grandmother would normally go, such as the small garden behind the house and the clothes drying area as well as the storeroom that we usually never went near. But she was nowhere to be seen.

On our first day here, we had seen Grandmother walking with a parasol on the path alongside the rice paddies. Recalling that, I went to the path in the nearby rice field on my way back after picking scallions I planned to mince and scatter over a tofu and ground beef sauce for supper.

While I was walking along the road next to which the green rice fields stretched far and wide, I heard a strange voice from somewhere. No, more than a voice, it was a song. There was a melody to it. But since the melody was not long enough to be a song, I thought it was after all better to call it a voice, as I looked around, puzzled.

Was a frog's cry a voice, or a song? The frogs were performing a great chorus. I stood in the rice fields, viewing the wide stretch of green all around me. In the voices that sailed over the thorny waves of the windblown green rice, I suddenly noticed that some human voices mingled.

Standing in the middle of waves of rice, I looked around curiously. Then I noticed a small house in the shade of a big camphor tree nestled in the rice paddies on my right. The human voices seemed to float out from there.

There seemed to be several people chanting in unison. Their voices resembled the cry of the frogs.

When I got close to the house, I finally figured out what those funny voices really were. They were chanting a Buddhist sutra. The sun was still high in the sky. The cicadas were also crying.

Wiping the sweat on my nose and forehead repeatedly with the back of my hand, I sneaked inside the hedge. I went around the back and found a verandah wide open. As I approached it, the sutra grew louder and louder. There was such resonance that the air seemed to tremble and vibrate. The cicada cries had disappeared, and only the chanting of the sutra overwhelmed the place with its great power.

I stuck my head out from the shade of the tree and found five or six old women perspiring as they recited the sutra. With their curved backs, they looked almost ridiculously tiny. Probably they sat with their bottoms flat on the tatami and feet slightly out to each side, instead of sitting with the rear end resting on the ankles. On their small bodies sat their little necks and wrinkled faces, and the mouths in those faces were opening and shutting simultaneously.

The old women's faces were flushed and steaming, and swollen into ruddiness. The high-pitched chanting had a strange rhythm.

Their chorus sounded like this to me:

> Gyaatei gyaatei gyate gyate
> Sowa sowa sowa gyaatei gyaatei
> Sowa sowa sowaka*

Again and again I wiped the perspiration on my nose with my hand. From among the vortex of scarcely human-sounding voices I picked out the voice of my grandmother. And from among the wrinkled heads that seemed to belong to nonhuman creatures I spotted my grandmother's head.

I felt very strange. Looking at my grandmother's face, I felt so indescribably weird. Both the appearances and the voices of those women were just like those of frogs.

Why did my grandmother chant in such a strange voice? Wasn't she embarrassed about producing such an inhuman voice?

But my grandmother was contributing her utmost to that vocalization. She was no longer her usual self. What strange beings old people were, I thought.

I was standing under a big camphor tree. From time to time a cool breeze drifted over to the shade cast by the tree and the eaves of the house. The perspiration seemed gradually to withdraw from my nose.

Dropping my eyes mindlessly on the moist black soil of the yard, I saw a colony of ants moving in a long line. The procession went from my feet to the back entrance and around the broom that leaned near the door, and disappeared into the corner of the house.

A group of ants that had branched off at the broom from the rest of the procession were climbing a rose plant near the hedge. The sight struck my eyes as somewhat startling. Between the thorns of the green rose stalk, the zigzagging line of ants advanced slowly.

The chanting of the old women ranged over the zigzag pattern of the ants. A few of the topmost ants eventually climbed all the way up the stalk and reached the center of a white rose.

That night—

This was the conversation of Shinjirō and Grandmother.

The two were sitting on the verandah facing out, eating watermelon.

"Tastes good, Grandmother," said Shinjirō.

---

*A slightly corrupt version of an incantation from *Hannya Shingyō* (The Heart Sutra).

Grandmother nodded silently.

"I can't eat watermelon if I die, can I?"

Grandmother still answered only by a nod of her head.

"It's good that that Suzujirō person didn't die, isn't it?"

". . ."

"He won't even be able to eat pineapple if he dies."

Shinjirō seemed to be leading Grandmother in a certain direction.

"Don't you want to eat pineapple, Grandmother?"

Grandmother shook her head for the first time.

"I don't want to eat any such thing."

Shinjirō's leading question was dodged very simply. He seemed indignant.

"But I want to eat pineapple, too."

Inside the room, Minako and I looked at each other involuntarily. Why was food the only thing my brother could talk about? Didn't he know any other way to open a conversation? Minako covered her face with both hands to avoid bursting into laughter, and looked down.

Grandmother said nothing, washing down watermelon. Shinjirō looked into her face.

"Grandmother, why do you dislike this Suzujirō person? Is there some reason?"

". . ."

"Like maybe you had a big fight with him a long time ago?"

". . ."

"Please, can't you tell me the reason?"

". . ."

Stealing glances at him, I somehow started to feel sorry for Shinjirō. He really seemed to want to go to Hawaii. I also felt sorry for Grandmother who was being pushed more and more toward Hawaii. In my eyes, the two formed a depressing painting that almost made me sigh.

"Say something," I heard Shinjirō say loudly.

"Shall we make him stop?" Minako whispered to me. The fact was that it was Minako who had asked Shinjirō to persuade Grandmother.

"Too late now," I said. Shinjirō might throw the watermelon slice in his hand. I kept nervously watching the backs of my brother and Grandmother.

After returning from the sutra chanting in the field, Grandmother had entered her room, saying nothing to Tateo when he asked where she had been. In her room, she seemed to be thinking intently about something, lonesome and lost.

Looking troubled, Tateo had said that this kind of family atmosphere was no good. In his view, it would ruin our precious summer vacation; besides, in the first place Grandmother had to put a stop to this kind of ambiguity before their parents came home. It was important because the parents would certainly try to decide how to handle the relationship with Clark's family, now that they had seen the situation over there. What Tateo had said was logical.

"Grandmother!" Shinjirō said in a loud voice. The watermelon was going to fly, I thought, and ducked my head. But his hand did not move. My ear caught his voice, now low and almost trembling:

"I want to go to Hawaii. . . ."

Minako and I bit our lips silently. He may have been crying.

"I'm sorry," Grandmother said, turning to Shinjirō. "I don't dislike this Suzujirō person or anything."

"Then why?"

"I can't remember him," Grandmother muttered, her head down. "I have no memory of such a person at all."

Shinjirō's back stiffened into silence. We waited and waited but he said nothing. He seemed to be staring at Grandmother until his eyes nearly bored into her face.

Grandmother, too, became silent.

Minako and I exchanged looks silently, and shook our heads. Both on the verandah and on this side of the sliding screens, there was a wordless hush.

Grandmother's brain showed no more symptoms of the illness that we had worried so much about after this incident. We had suspected that she might be suffering from the forgetfulness of old people or something like that. Considering her age, it had even seemed natural that she would succumb to such a condition.

After that, we gathered in the dining room as if we were aware of nothing, and, pretending that we were begging her for tales of the old days, asked her about people we didn't know. Grandmother came from a family of thirteen children. She was one of only several still striving to live in their old age. Although Suzujirō had just been added, that didn't change the balance: there were still fewer survivors than dead.

Tateo counted with his fingers as Grandmother mentioned her brothers' and sisters' names. The number was twelve. Since she counted herself, it followed that her memory skipped just Suzujirō.

"There were thirteen of them," Tateo muttered. "It may be possible to forget one or two. Think of me: if I'm told to recall all the neighborhood friends I played with in my childhood, it'd be impossible."

It may be so, I thought. Siblings as they were, with as many as thirteen the age differences would be wide. A very much younger or older sibling might have felt more distant than a neighborhood friend.

Suppose an eighty-year-old grandmother recalled the days when she was fifteen or sixteen. That would mean recalling the past of over sixty years ago. When we searched our memories, we could only trace back ten years. Even Tateo, the oldest of us, could remember back fourteen or fifteen years at most. We had no idea what it was to trace sixty years into the past.

What kind of structure were memories stored in once they were accumulated in the brain? Far in its recesses that belonged to the past, and probably sunken at the bottom of something like mist, what shape did memories of the world sixty years ago assume? Among Grandmother's siblings bobbing and sinking in her memory, I thought, was the boy Jikurō. The old man called Suzujirō, in other words, had sunk down with no knowing when he would float to the surface again.

Grandmother listed the twelve names with unexpected firmness.

"Ryōtarō, Kōjirō, Kōichirō, Tetsurō, Shōsuke, Senji, Jikurō."*

Tateo counted: one, two, three, four . . . "Impressive, Grandmother," he exclaimed.

"Ginrō, Shūrō, Tsumisuke. How many now?"

"It's ten. Two more besides yourself."

"Then, since the other two are girls, it's me and my younger sister O-Mugi."

"Great," Tateo sighed. "You'd have passed into college at the first try."

The number she came up with was twelve; the remaining one was poor Suzujirō.

Grandmother then told us about Tetsurō, the seventh boy, and Tsumisuke, the eighth.

"Tetsurō had a handsome face and carried himself well. Every time he went into town, he'd come home with a letter in a female hand, folded small and tucked into his kimono sleeve. He couldn't bring it home, so he went round the wall of the estate to the bamboo grove in back. After opening and reading it there, he tore it up and lighted it

---

*Each of the names of Grandmother's brothers, some of which are unusual, contains a kanji with mineral or geographical associations such as "diamond shape," "mine," "balance," and "to bore." The name of the brother in Hawaii, Suzujirō, contains kanji for "tin"; Tetsurō the shoemaker has kanji for "iron"; and Jikurō, who wrote kanji with ink and brush, has kanji for "axis" (which also means brush holder or calligraphy scroll). Their sisters' names have rural associations: Grandmother's name Sanae means "rice sprouts" (her married name means "flower mountain"), and her younger sister's, "wheat."

with a match. So, there used to be little, burnt pieces of paper, scattered here and there in the bamboo grove."

"He was quite popular." Tateo looked slightly envious.

"One day, somehow he suddenly said he wanted to be a shoemaker. He looked pathetic as he confided this wish to our father, but his reasons weren't clear. He said only that he wanted to become a shoemaker so please let him leave home."

Grandmother looked as if at a loss.

"Father was really surprised, and asked why again and again. But Tetsurō said nothing. All the other male children in our family became teachers or priests, or farmed, and no one else said anything like that."

"Did your father let him go?" Shinjirō asked.

"No. Tetsurō just left on his own, bypassing Father's disapproval. He went down the mountain, and became an apprentice at a shoe store called . . . something, in a nearby town."

Having told us this much, Grandmother lowered her small shoulders and sighed.

"Once my father went into town and peered through the display window of that shoe store. Tetsurō was apparently working by the side of the keg-shaped stove with someone who seemed to be the master. He was busily tanning leather on a rag spread over his lap. Father kept staring at him. And then he came home, looking terribly lonesome."

"So he did become a shoemaker," said Minako.

"He trained in the master's shop for five years. Just when he seemed ready to start off on his own, he got into trouble."

"I know. A woman, right?"

"That's it, that's it."

Facing Tateo, Grandmother nodded many times. I felt that the story was developing in a direction that was not good for Shinjirō, who was leaning his back against a pillar as he listened.

"Why don't you go to bed?" I said to him in a small voice.

"I'm not sleepy," he answered, looking at me as if puzzled. Tateo and Grandmother after all were different in this sort of thing from our parents.

"Tetsurō ran away holding hands with the master's wife. It seems she was the object from the beginning."

"Grandmother, wasn't the master's wife old?" Shinjirō suddenly asked. I gave up the idea of taking him off to bed.

"She was his second wife, and still young. From what we heard later, the two ran away along the mountain trail on a September night, without even a change of clothes. Running away with someone else's wife would have been awful enough, but he stole the wife of the master he owed so much. Torches were burned and the voices of many people came up from the foot of the mountain. The woman collapsed on the way. They told me it was the most frightening thing they ever experienced."

"Did he run home?"

"He came close by, but he was afraid that they might get caught, so they went up another, even higher mountain, and trekking along the ridge, they settled as a couple far away."

"A terrific story," Tateo said, shaking his head. He breathed a few times, repeatedly, then fell on his back on the tatami. "How amazing that a man like that existed before us—such a great adventure of peas!" he said, with his eyes closed.

"I had already married into this house around that time. Tetsurō and the woman went across the mountain just above us," Grandmother said, looking around at our silent faces. "And when they were walking along the trail through the cedar grove stretching all the way to the mountain on the other side, they happened to see a reddish brown spot in the middle of the beautiful deep green grove. The moment Tetsurō glanced at it, he realized that some trees must have been burnt dry by the lightning. Two cedar trees were standing, reddish brown in the middle of the green grove."

We all gazed intently at Grandmother's face.

"Then the woman walking by Tetsurō's side, looking at those cedars, said, 'Look, those cedars look as if they commited love suicide.' "

Tateo remained flat on the tatami.

"So, the two crossed the mountains and set up house," Grandmother said, concluding her tale.

After this came a sequel from Grandmother's mouth that nearly glued Tateo to the floor.

"Maybe it was divine retribution. Over a dozen years later, when Tetsurō was now a master and had a young disciple in his house, this evil fellow tried to run away with some money. Tetsurō caught him and the two wrestled. And Tetsurō died, when his disciple hit him with a metal hammer."

Tateo was silent. The romance of the peas ended with fearsome results.

The old master's wife who had become Tetsurō's wife died of illness soon afterwards. They left behind one pea, a boy.

"Tetsurō's younger brother Tsumisuke, who had no children, adopted and raised this baby as his own."

Grandmother's father had as many as thirteen children, but no one else died the way Tetsurō did. Her father never left home after that.

Grandmother's story turned out to be long, and sounded like a distant tale, an indefinable wind in a dream within a dream. Suddenly I noticed that Tateo, still on his back, had turned pale.

From his uncharacteristically serious expression, I recalled the story he had once told me:

"My old man apparently isn't Grandmother's son, but a child of her younger brother, the second youngest or so in the family."

Tateo had looked happy, though I didn't know why, when he said this.

"Do you know the term lineage legend, Tami? When I heard about this I got really excited. I wonder what kind of person my father's father was."

But now Tateo was pale. It would seem that Tetsurō was that very man.

Before my eyes, the pea that had proceeded from the second pea which, in its turn, had come from the original pea, was lying completely lost on the tatami, seemingly unable to utter a word.

III

Once every three or four days, I went shopping in the town at the foot of the mountain, riding on the rear rack of my brother Shinjirō's bicycle.

We went down the mountain on a road between cedar groves which were gloomy even during the day, and looped around and around through a big bamboo grove. Except for occasionally passing a villager's car, we encountered no one.

Then how did Grandmother, living alone, get her food? A car would come up from the town below, loaded with meat, fish, and daily miscellanies, and set up shop. It would go around vending door to door, but I preferred to go shopping on the back of Shinjirō's bicycle.

The bicycle, quite ancient and old-fashioned, hardly seemed of any use when I found it in the storeroom. Shinjirō was an odd perfectionist. He set about putting air in the tires, changed the chain, went to buy wax for taking off the rust, and polished the body. I went to the backyard every day to watch his progress.

From time to time the sound of Tateo's organ came all the way to the backyard. Shinjirō was oiling the bicycle using the machine oil he had taken from a drawer of Grandmother's ancient black Happy Machine.* Seeing me crouched down and peering closely at the bicycle, he solemnly promised,

---

*A brand of sewing machine.

"Tomorrow for sure, I'll take you for a ride, I'm positive."

When Tateo got bored of the organ and started taking a nap on the verandah, Shinjirō was taking the rust off the front wheel of the bicycle. Having woken from his nap, Tateo was drinking tea in the kitchen, when Shinjirō finally finished polishing both wheels and went on to the pedals.

He remained squatting near the bicycle, perspiring, until I went to announce that dinner was ready.

"Done!" he shouted, rising. He immediately hopped on the bicycle and rode it from the morning glories in front of the clothes rack, and back around the pond.

"Get on," Shinjirō said as he put down one foot, tilting the bicycle.

The rack for me to sit on was sparkling from corner to corner, the dust and rust gone. The next morning, Tateo took a look, and was saying something to Shinjirō. Tateo got on the bicycle, seeming to have gotten Shinjirō's permission.

"Hey, this is in pretty good shape. Not a bad job at all," he commented from the bicycle as he pedaled smoothly.

This was perhaps the first time I got annoyed at Tateo. Tateo was the kind of person who tried to get things done, if possible, without any effort on his part; true to his nature, he was too easygoing and had no hangups. He rarely got angry. I didn't dislike this oddly shallow aspect of him. I got upset for Shinjirō.

"You fool," I glared at Shinjirō, moving near him and commanded, "tell him to get off."

Tateo laughed as he got off the bicycle.

No one, including Tateo and Minako, got to ride the bicycle after that. Shinjirō diligently carried me down the mountain. On the way back, we both walked: we put our purchases on the front and back and pushed the bicycle up the hill. This was much easier than carrying things by hand.

Since we came home along the cedar path near the creek, this meant a half-day shopping trip. But we got home late not only due to the distance, but also because of the fun we had on the way home.

Coming back from shopping one day, we decided that we might be able to find the waterfall if we traced the creek to the upper stream. We recalled that Grandmother had from time to time mentioned going to the falls with the laundry, partly for fun.

"She went there with her laundry, it can't be so far," Shinjirō said. "I think it's just that the shortcut got covered with grass because no one passes that way now. So let's try walking along the path, and listen real hard."

We climbed back toward the village, Shinjirō pushing the bicycle, and I eyeing the tall grass on the side of the path. About halfway up, I started walking the bicycle while Shinjirō pushed his way through the grass. The sun was shining right overhead, whitish and almost melting, but it was not so hot because of the mountain air we were steeped in. A bird's chirp came from an unexpected place.

We had climbed a fair distance when Shinjirō shouted, "Yay!" He was buried in the tall grass, but I always knew where he was because of the waves he made as he moved, like the periscope of a submarine.

Soon he popped out, noisily shoving aside the waves of grass. "Listen hard, you can hear the fall," he said, drawing me into the grass.

"I hear it." I nodded.

"Right," Shinjirō said.

In the grass we stayed very close. We would have lost sight of each other if we got separated.

"Stay where you are, Sis. I'll be back after I arrange the things on the bike," Shinjirō said, and bustled out to the trail. I was just beginning to feel helpless from staring up at the sky from the bottom of the waves of grass—it was perhaps more than ten minutes—when he finally returned. Then he led the way, leading me by my hand. Though we couldn't see the ground, it felt like we were going further and further down a fairly steep hill.

After descending for a while, the grass ended and we came out to the old shortcut. By then we could hardly hear each other, surrounded by the sound of the waterfall. At the bottom of the little path, we came face-to-face with the fall. We had to crane our necks way up to see the top of the fall. It was a dark fall shaded from the sun.

". . . ." Shinjirō said something.

I brought my ear closer to him and yelled, "WHA-A-T?"

"THERE'S A SNAKE, SO . . ." Stay still, he was going to say, but before I heard him out, I jumped with a scream that nearly made him trip over himself. Probably I kept screaming for a minute or so, then leapt, my eyes on the sky because I was too scared to look down. In the meantime the snake went down to the water and swam away. By the time I looked where Shinjirō pointed with his finger, it was already sliding in a long S shape on the surface of the dark water near the base of the waterfall.

This was why I disgusted him by repeating, "Let's get back, let's get back," when he suggested that we stay longer. I kept looking restlessly at the ground, hardly even glancing at the fall. I heard it in its fullness, however, for the rumbling tumbling noise of the water overwhelmed my shaking heart with even more power than before.

Cutting through the thicket, we came out to the village path, as the sound of the fall grew distant from our ears. I finally recovered my spirits.

Shinjirō passed by me and climbed a chestnut tree in front of us. Lowering one shopping bag after another, he handed each one down to me.

"Just in case a fox might come out and eat the food," he said, to explain why he had hung the bags.

Tateo seemed strangely quiet, Minako observed.

Every day he napped, whether on the verandah, in the little back room where the organ sat, or in the shade of the flowering paulownia in the yard. On the surface, his attitude had not at all changed. From time to time he offered two or three long yawns into the pages of the book he was reading. Then, for example, if he was in the yard, he reached out to put a flower in his mouth.

When I was small, my mother scolded me for copying Tateo and eating an azalea blossom. The reason she gave was this: "Only bad boys eat such things." Although I didn't understand what was so bad about eating flowers, I somehow felt she wasn't entirely wrong.

Man that he was, Tateo's lips were pale red and attractive. They were fit for eating flowers. On the other hand, I seemed to be a girl whose lips never looked just right, even if I held a flower in my mouth.

It appeared to be a fact that something began to change in the mind of Tateo who, however, looked rather unchanged on the surface. He might have been pondering the misfortune of his real grandfather whom he had come to know about through Grandmother's story.

That day Tateo lay on his back on the tatami floor of the organ room, feet propped high against the wall and eyes open. On the wall hung framed pictures of school outings from the time when she taught at elementary school, the award for working decades without a break, and so on. One of those framed documents that had turned yellow was a speech for elementary school children of the old days. In the upper right corner of the paper was the oval photograph of a man's face.

The caption read So-and-So, President of Tokyo Imperial University. That meant he was an important professor. The article carried the old, old words of this teacher of a university Tateo could never ever enter even if he stood on his head.

"Have you read this, Tateo?" I asked, approaching the wall to take a look.

"Yes—" Tateo answered.

It ought to have caught his eye, the way he killed time there every day.

I read out loud in front of the wall.

Form Good Habits

Although bad habits are easily acquired, it is quite difficult to cultivate good habits. It is praiseworthy to discipline oneself in spite of this. Rise early in the morning, avoid likes and dislikes in food, help your parents at home and go on errands for them, review your studies thoroughly every day—there are many other good habits.

From long ago, praiseworthy people, at your age, mustered their courage to cultivate good habits.

Children, if you strive to form good habits, even one by one, you will internalize them, and before you realize it, you will become exemplary Japanese.

Tateo silently listened to me read. Then he said, "That was so bad," incredulous at my incompetence in handling the kana spelling system no longer used.

"Fine with me," I retorted.

At least I had formed the good habits described here. I might be the ideal girl the teacher of this article had commended. Tateo was the opposite: a boy who was full of bad habits.

He should do some soul searching on the example of his grandfather who had been hit with a hammer.

"You know, Tami," Tateo said suddenly raising his body. "The cedar trees Grandmother talked about the other night are still supposed to be on the mountain above us."

"Cedar trees?"

"The two cedars that committed love suicide when it thundered," Tateo said, his eyes bright.

"Come on. How could dead trees from such a long time ago still be there?"

"But pine trees stay standing," Tateo said. Then he peered into my face and whispered: "Shall we go and see if it's true?"

Something that had seemed a tale of a far, faraway dream was beside me even now. I fell silent for a moment.

Tateo stared at me as he waited for an answer.

"It's a long walk, isn't it?" I asked.

"There's the bike."

What crossed my mind at that moment was Shinjirō's face. Somehow I felt sorry for him.

"Let's go see. They say that lightning leaves zigzag scars on the trunk of trees, as if a cat clawed at it," Tateo said, rising.

What loomed next in my mind was the back of Shinjirō as he had gone out a while ago. Probably he had gone to the temple to take a nap. If so, he wouldn't return until suppertime.

"What about Mina?" I asked.

"Idiot. How on earth do you expect to get three people on one bicycle?" Tateo was right.

We climbed along the mountain trail, passing not a soul.

The trail was wide, the slope mild, but as he pedaled, a map of sweat soon formed on Tateo's back.

"Hold on tight," Tateo said.

"You're sweaty, so . . ."

"So what?"

"So it feels gross."

"You're getting a free ride, don't be picky," Tateo said gaily.

We curved slowly to the right, then again slowly to the left. The long slope of the cedar grove came in sight. Beyond that the sky was like a blue depth.

Had Tetsurō and the shoemaker's wife really passed such a place? Grandmother's story had been frightening, but this mountain wasn't at all scary. We heard a cuckoo from the direction of the lake behind the mountain.

No matter how far we went, we saw nothing but deep green cone-shaped outline after outline of cedar, puffing up and out into each other. The sky was still like a hollow; the mountain was soundless as though it were a paper model. Shinjirō's well-polished bike raced along.

About the time the slope gradually grew steeper, we suddenly spotted a strange color amid the cedar's deep green on the right. Dry, reddish brown wood was stuck in the middle of green trees, as if a spot had faded in a photograph. I strained my eyes and found that it was two cedars.

"That's it," Tateo said as he stopped the bicycle. We got off and started to walk toward the grove. Close up, however, the thick cedar grove simply towered sky-high, intercepting the sun's light. We gave up the idea of entering the dark grove.

"It's good enough to view it from here," I said to the disappointed Tateo.

"Well, I guess we might lose sight of them if we go inside," he responded.

Tateo and I sat on the edge of the trail. It was so quiet that my ears almost hurt. Chi, chi, chi, a bird flew by. After the bird passed by, it was hushed again. In that stillness Tateo whispered:

"Those are different."

"What's different about them?"

"They aren't the cedars in Grandmother's story."

"Why?" I asked, staring at the cedars on the slope, "The two of them are dead together, just as they are supposed to be."

"Well, it thunders a lot in a high mountain like this. Occasionally two trees can get hit on the side and go at the same time."

Tateo shook his head, and pointed at the trees:

"Take a good look. How could cedars that died decades ago still have any needles? They should have fallen off long ago and left skeletons behind."

Tateo must be right, I thought. Like a woman's hair, long brown branches and needles were hanging thinly here and there on those cedars.

"I wonder if Grandmother thinks those are the trees that Tetsurō saw."

"Probably. We don't see any other cedars burnt by lightning."

"What happens when these cedars rot?"

"Two other cedars will commit love suicide," Tateo laughed. It wasn't his usual low voice. My eyes focused on the brown cedar needles. The long needles of the two trees, or the strands of hair, looked intertwined.

Chi, chi, chi, chi, a bird passed again. The forest became quiet again. The silence brought back a vivid image of the dead. The interlocked strands of hair made the two dead trees look as if they were talking to each other. I felt as if my ear caught a terrifying dialogue.

The wind blew. The branches of the live cedars, rich with green needles, trembled. It was as if they breathed in the wind. The cedars of love suicide, however, only faintly shook their reddish brown strands of hanging hair.

Until then, I had portrayed death in my mind as wearing a garment that was light as a whiff of wind and free as a butterfly. However, the cedars of love suicide appeared very heavy. They seemed to bear an invisible weight, and I felt suffocated.

"I'm going home," I said to Tateo.

"Why?" Tateo, who was standing close to me, looked searchingly into my face. He looked even deeper into my eyes and into my trembling heart, as he asked once again, "Why is it . . . ?" It didn't sound like his voice.

"Tami," Tateo said with a strange emphasis in his voice. Then, abruptly, he hugged my shoulders tightly.

"Whoa," I shouted loudly. Surprised by the pitch of my voice, Tateo let go, but I was still in shock. I slapped his hands again and again without reason. In my mind I saw the strands of hair on the love-suicide cedars slipping down toward me.

Behind me I heard Tateo calling. Without turning back, however, I ran frantically to the bicycle at the edge of the trail, and jumping on it, biked down the mountain as fast as I could.

I went home, leaving the trees of the love suicide and my degenerate cousin.

It was very quiet that night.

Tateo seemed to have finally reached home on his exhausted feet late in the afternoon. Minako came by to tell me as I was preparing supper. She said that as he washed his feet alone by the well, he kept sighing.

I prepared fish boiled with soy sauce and farm lotus marinated in vinegar and sugar. Tateo hated fish. He liked sour things even less. Mixing the thinly sliced lotus in generous amounts of vinegar, I wondered if Tateo would have a fit. I had purposely left out the sugar for his portion.

His expression as he ate was so ghastly, it seemed his face might collapse at any moment. He had to eat because I had said to him in front of everybody, "I don't want to see any leftovers."

Upon finishing his fish and the pickled lotus, he drank many cups of tea and withdrew to his room.

After dark, Grandmother asked Tateo to write a letter to Hawaii for her. It was to be addressed to Suzujirō. Tateo came to the dining room with letter paper and a fountain pen.

"Greetings," Grandmother said in a formal manner, sitting in front of Tateo.

"Greetings," Tateo repeated, his fountain pen racing.

"Since our last exchange, how has your physical condition been, if I may inquire."

"Since our last exchange, how has your physical condition—physical condition? wow—been, if I may inquire. Tami, how do you write the kanji for 'inquire' as in inquisition?"

With my back to Tateo, and eating pears with Shinjirō and Minako, I said I didn't know. Shinjirō swallowed the big chunk of pear in his mouth.

"Write 'yo' in katakana, 'e' and 'ro' under that. Then write 'sun' in kanji underneath," he answered proudly.

"My, how smart Shinjirō is, what can I say?" Minako mimicked Grandmother's tone.

"So that you will thoroughly recuperate as swiftly as possible, I am offering prayers day and night to Buddha and the gods."

"Grandmother's a liar," Shinjirō said.

"She's praying for him in her heart," I said.

"It's too noisy, be quiet," Tateo said. "Let's see. So that you will thoroughly recuperate as swiftly as possible, I am offering prayers day and night to Buddha and the gods."

"Thanks to the sun," Grandmother continued with a serious expression.

"The kanji for 'thanks' is the grass radical and 'yin' as in 'yin and yang,' " she added in a tone she must have used as an elementary teacher long ago.

I always found it strange that Grandmother's memory was very uneven. She remembered specific details with surprising clarity. However, this was always limited to a small part of the whole. In other words, her memory demonstrated power only where it was not particularly helpful.

If she had forgotten the kanji for 'thanks' and instead even vaguely remembered her Hawaiian brother called Suzujirō, that would help people around her in so many more ways. Suzujirō might get better more quickly, too.

"Now let me beg of you," Grandmother said.

"Let me beg of you."

"Years and months having gone by through many a star and a frost as the brothers and sisters once of tender ages have traveled far on the tide of time, today most of my memories have vanished like phantoms, and—"

Tateo became very busy. Moving his fountain pen quickly, he repeated after Grandmother.

"Your countenance and appearance, too, have gone to oblivion."

"Your countenance and appearance, too, have gone to oblivion."

"Sis, what's Grandmother saying?" Shinjirō asked.

"Shhh. Be quiet," I said.

"Please be so kind as to send me one or two photographs of yourself, I beg of you."

"Please be so kind as to send me one or two photographs of yourself, I beg of you."

"To Mr. Suzujirō Haruno, with respect."

"To Mr. Suzujirō Haruno, with respect."

Grandmother let out a sigh.

Sitting with her back bent, she tilted her deeply etched, small face downward. Somehow she looked lonesome.

When the letter was finished, it became quiet in the room.

Shinjirō and Minako faced each other, eating pears.

Tateo and I sat back to back, ignoring each other.

Outside, the pale moonlight came filtering down.

The cedar grove as it must be on a moonlit night surfaced in my mind's eye.

The cone-shaped cedars stood, shining as though bathed in snow.

IV

The menu for that day was rice pilaf with small clams. Since Grandmother had never heard of it, I renamed it to sound more traditional, "Western-style rice with shellfish."

"Oh, oh, rice with clams in their shells," Grandmother sounded as if startled out of her senses.

"My, what can I say, this looks like a meal for pirates!" Grandmother opened her mouth like a sea shell as she lifted the lid.

"The shells open up in the pot and release tasty juices," I explained.

"Then you're supposed to eat the rice and at the same time pick up a shell, so you enjoy both together, right? My, this is going to be a busy feast," Grandmother said. Then holding a spoon in her left hand and picking up a shell in her right hand, she opened her wrinkled mouth like a coin pouch tied with a string, sucked the juice in the shell, then sucked in the clam, and finally licked the grains of rice stuck to the inside of the shell.

"Delicious, delicious," she said. "Thanks to my granddaughter I'm here enjoying such a meal, like digging shellfish on the beach—it's so lucky I've lived this long."

Amused by her words, we kept laughing, our spoons shaking.

Dinner took longer than usual, with Grandmother slurping her way through the entire time.

Washing the dishes being her chore, Minako stood in front of the sink. The bath water was Shinjirō's duty, so he went to the bath on the other side of the kitchen. Tateo, who had no job, simply remained seated with Grandmother, supporting his belly with satisfaction.

The old plates with their mismatched dharma, eggplant, and cucumber designs were piled up in the sink, then washed and stored away in the cupboard as before. In the meantime, the bath was heating up.

Grandmother and I took a bath.

Tateo and Shinjirō had bathed first. It was Minako's turn to clean the bath, so she was bathing last that day.

The cries of crickets, reaching us through the screen window, were ringing in the tiled room. Grandmother's back was in front of my eyes. As I rubbed her skinny back with a towel, I recalled the Mt. Fuji game that I used to play with Shinjirō when we were little.

Firmly pinch and pull up the skin of the back of your hand. Because the flesh is thin, the skin pulled up makes a neatly defined mountain. Shinjirō's mountain was small, and mine was a little bigger. While washing Grandmother's back, I recalled those mountains. Given her skin, a fine Fuji would form no matter where on her body. All skin—when I saw her from this point of view, her body was a really pitiful sight. However, looked at from the angle of making Fujis, her body strangely took on a special value.

As I was thinking of the flesh on her arm, then her leg, my eyes were suddenly attracted to her back. Given this area, an extra large, spectacular Fuji could be formed, far better than the ones on the back of a hand.

I placed an imaginary Mt. Fuji on Grandmother's back. I could see the deep creases and their shadows that ran from the mountain top down to the foot, as well as the silhouette of its wide, wide skirt. Under my pinching fingers, the skin of Grandmother's back would beautifully form into that grand piece of nature.

Fighting against the temptation to pull up the skin, I rubbed her back yet harder.

Grandmother turned to me in the steam. It was my turn to let her rub my back.

"Oh . . . , so beautiful. How beautiful your back looks. It's so glossy," Grandmother said. Since I didn't like my oily skin, I wasn't at all pleased by such words of praise.

"See, it repels the water. What a beautiful sight." She always followed this with her habitual expression, "Heavenly, heavenly."

Above the glittering bath water, two heads floated, Grandmother's and mine. From sitting in the hot tub, her face was now red, and reminded me of a walnut soaked in warm water. And by the side of her head, her favorite paradise was gently bobbing up and down.

Looking at that paradise, I thought: Grandmother has perhaps always lived together with paradise in this way. Then, half of her heart was already taken away to that side, and things from dozens of years ago, to her, were like a dream.

No, even this bath we were sitting in might be half in her dream, and her granddaughter might be her newest dream— Then . . . then, what did the world look like in the eyes of a person when only half of her mind was awake and working?

I was thinking of such things as I faced Grandmother's head, but in fact the waking half of her mind was opening the door of an old memory, with me as its threshold. She was gazing happily at my face, but then she let a grave matter slip:

"Oh, the way you carry yourself, Tami, you're so much like your mother," she sighed in the tub. "Really how very much like Mugiko—"

"What did you say, Grandmother?" I asked. "Who's Mugiko? Why do I have to look like Mugiko, or whatever her name is?"

Grandmother's expression changed visibly. She looked alien to me.

"Who's Mugiko, Grandmother?" Over the surface of the bath water, my head glided toward her.

Grandmother was at a loss for words.

Then gradually her eyes, eyebrows, and mouth began to collapse. Behind the skin of her face, too, it was clear to my eyes that something was collapsing in the same way. Slowly, and it seemed at once long and short, tears filled her eyes, and she started to cry with her hands over her face.

"Is she . . . ," I asked, pressing my beating chest, "can she be my real mother?"

Grandmother finally nodded.

"She's different from Mom, isn't she?"

By now Grandmother was simply nodding.

I was sitting in the water then, but abruptly felt a sensation of losing my balance. It was as though I had lost my legs. My legs, the crucial roots that supported my body, were suddenly pulled away. It flashed through my mind that parents and blood relatives were like legs. Without them, I thought, one swayed insecurely and became helpless. . . .

I was deeply struck by this clear sense of reality. But I couldn't just sit there and continue to be amazed by this refreshing discovery. I realized that I had to ask Grandmother about Shinjirō.

"Shinjirō's a child of your current parents," Grandmother answered in a subdued manner.

To be sure, we were sister and brother who were totally different in face and physique. I became silent.

Dizzy from the long soak, Grandmother tottered out of the tub to the washing area. I followed her. Both of us flopped onto the tile floor and just sat there.

In front of my eyes it grew alternately dark and clear. Grandmother's face, so much like a walnut monster, now emerged and now disappeared. Probably my face also looked that way to her. This symptom of dizziness was caused by the bath, but my heart was in the same condition. In a word, it was all dark in front of my eyes.

At the lower end of the steam, Grandmother started to cry, a wet towel pressed to her face. Then she started to talk:

"Mugiko was my younger sister's daughter."

Now I remembered: the name of Grandmother's younger sister was O-Mugi. Mugiko must be O-Mugi's daughter.

"Mugiko died right after she gave birth to you," Grandmother cried for a while again.

"At our place my son Shōsuke had just married Sumiko. We agreed that it was the only way for the baby, and decided to ask them to take you."

Shōsuke was my present father.

"People cried with joy at my sister's place—they were so grateful to Shōsuke. So right after birth the baby became his and we had her registered under his name at city hall."

"So Mugiko died right after I was born?"

"I think it was a day or two later."

"So the person who was my father didn't say anything? He was giving his baby up to someone else. He said nothing?"

"Your father . . . ," Grandmother said. As if in answer to the question, tears trickled down again.

"He died, right?" I asked.

Grandmother nodded. People in the old days seemed to have died so easily.

I recalled having imagined a sad story like this once. It had been at night, and I had enjoyed seeing the fantasy unfold in my mind's eye as I lay in bed. It was not just me; most girls got thrills from that kind of imagination at least once or twice. Minako, for example, even seemed to hope to become an orphan. We craved so much for the romance of a green pea, the sad pea that Tateo talked about.

I thought that if it were possible, I would like to convey to Minako the feeling of a girl whose fantasy had become reality. Embracing a feeling that was partly regretful, partly sad, partly patronizing, and partly lonesome, I looked at Grandmother's naked figure.

What unfortunate people there were in the world. I thought about my real mother, Mugiko, and a single tear that felt false rolled down. There had also been, I thought, a married man who died before seeing his baby's face. . . . Here another tear fell.

And . . . and then there had been a baby who, though lucky enough to be born, was without father or mother. Yet another tear fell. That unlucky baby was myself.

How false this real story sounded. I was deeply moved by this thought as I cried.

"Reality . . . ," I thought, toweling my face that was wet with tears, "even if it doesn't look like the truth, is after all reality."

I spent the following days in profound melancholy. Now, trading places with Tateo, I was the one who sighed often under the shade of the paulownia. No matter what objects I observed around me, they were reflected in my eyes as if through a sheet of pale glass. Did every orphan in the world, I wondered, wear the same glasses as mine?

I want to talk about my mother's ribbon knots.

My mother, or my kind foster mother as I had now discovered her to be, had from girlhood been unable to tie her hair ribbons properly. When she put her hair up in a ribbon, her butterfly spread its wings vertically like a figure eight. Ordinarily a girl like that would ask her mother to tie it for her, but her mother also made vertical butterflies.

So, who should tie her ribbon every day before going to work but her father.

And the daughter of this clumsy mother also had to have her father tie the strings of her apron.

"What a helpless pair," my father would shake his head, looking at my mother and me alternately. Even Shinjirō could tie apron strings, yet we, his mother and her daughter, couldn't. In passing, since both mother and I could decently sew either by machine or by hand, this didn't seem to be a question of whether we were clumsy or deft.

Then, what did our vertical bow knot derive from?

When mother and I visited her parents' home, there was the hilarious sight of three vertical apron knots behind us as we stood side by side facing the sink—in other words it was impossible not to think that the vertical bows of the grandmother, daughter, and granddaughter were nothing but a result of what Tateo had once called cross fertilization of peas.

However, the peas were smashed by Grandmother's story. They were not cross-fertilized. . . . This was the fact.

I recalled with affection the face of my mother, who was in Hawaii and whom I had not seen for a while. Too bad for Mugiko, my natural mother, but, thinking of my mother's vertical bow knot, I shed nostalgic tears under the paulownia.

Shinjirō noticed nothing about my change. Since I wished to remain his big sister even now after the fact had come through to me, it soothed me to know that he sensed nothing.

Under the paulownia, I renewed in my mind's eye the memory of the days when I had gotten him to correct the knot of my apron. Not just the apron strings; in junior high school days he tied—no, I made him tie—the ribbons for my braids.

I was the kind of big sister who had many devices for making him listen to me.

I recalled with keen nostalgia the scene in which his boyishly thin and somewhat bony fingers moved slowly before my eyes. His eyes were on my hair, while my eyes were focused on his fingertips. The color of the ribbon was navy, as dictated by school rules.

"Is this okay? It's okay, isn't it? Come on, let me go. I'll be late for school."

I leaned my back with a sigh against the trunk of the paulownia.

From now on I would become a gentle, loving big sister, I thought.

One afternoon, Shinjirō brought Minako to the shade of the paulownia.

"Let's go to the waterfall," he invited. "Mina says she wants to see it once."

In a flash, I remembered the snake—that long S shape swimming on the surface of the water. The moment I thought of it, I had goose bumps.

"Shin says he'll take us there."

Minako wore a white hat to shade her face against the sun. I realized that Shinjirō and Minako would go, even if I didn't. I involuntarily stood up.

And I said to Shinjirō who stood smiling at Minako's side, "Go get my hat!" He clucked his tongue and ran toward the house. I had a tendency to ruin resolutions within minutes of making them. Shinjirō came rushing back, the hat fluttering in his hand.

The three of us went down the mountain trail on foot.

Minako and I walked side by side, Shinjirō leading us. Our landmark was the tree on which he had hung the shopping bags last time. His eyes darting, examining the shapes of trees, he finally spotted the right one.

"Not even one electric pole, only trees and grass grow here," Shinjirō tried to excuse his slowness to Minako.

The three of us held hands and noisily made our way down through the grass. My mind was already full of anxiety about snakes.

"Shinjirō, tell us quick if you see a snake." The moment I unthinkingly said this, Minako started getting scared.

"Snakes come out here?"

"They do. Really long ones," I said, with a thrill of pleasure.

"Shin didn't tell me that," Minako said, suddenly still in the grass.

"A little one wriggled out, that's all."

"So there was one! It really came out!"

As I watched Minako's face, I felt better and better.

"It came wriggling out toward my feet. It was blue with grayish spots."

"That's a lie. There's no such thing as a spotted snake," Shinjirō said.

"But I saw it. It came right near my feet."

Minako started to tremble.

"We've come all this way, we can't go back just like that," Shinjirō said.

"Let's go on. It's all right. If I find one, I'll throw a rock," I said.

Strangely, I felt I really could do it. As long as Minako was scared, I felt I could refrain from my own fear of snakes. I had become a rather mean girl.

Climbing all the way down the thicket, we came upon the fall. We seemed to have swerved as we went down, for the fall that faced us last time was now a long way to our left.

Leading Minako by the hand, I crossed the rocks that were wet with the spray from the fall. Last time, due to the fuss over the snake, we hadn't even gotten near the fall. Minako and I crouched down on a big, flat rock.

"Where did the snake come out from?" Minako asked in a loud voice. She looked flustered.

"Over there," I pointed. Minako followed my finger with her eyes wide open. I looked at her soft hair and her doll-like white profile that never got tan. Then I thought about my own tough, black hair and dark face that wasn't at all girlish. Probably that unlucky woman called Mugiko, like me, wasn't at all beautiful. There was something unfair about it all.

We were in a big shady spot.

Foliage hung over the fall from the cliff.

The spray, turning into fog, danced through the entire rocky area.

"It's gotten a little cold," said Shinjirō, who came near us hopping from rock to rock. The hair on our arms rose.

It was on our way back that the snake appeared. We had left the fall, hopping on rocks to return to where we had come down. The snake, slipping out from the grass, cut off Minako and me.

Minako did not scream. She just stiffened silently, grabbing on to my arm without a word.

"It's all right," I nodded. Yet, I don't know what was all right, for we remained standing immobile. The snake was of a color that could not be defined in one word. It looked blue, but sort of blackish, and also maybe gray. There were no spots anywhere.

That eerie compound color of blue, black, and gray moved slimily. I shuddered. And I saw as if from afar that my foot kicked a small pebble, aiming at the snake. The letter S turned and again turned as it hit the dirt with its tail and glided over a rock, until, shining and glistening, it disappeared into the grass.

"What's up?" Shinjirō's head stretched above the grass.

"It's Tami," Minako shouted in an excited voice. "Tami threw a pebble at a snake. It hit the snake."

Shinjirō burst into loud laughter. "Give me a break," he said and laughed again.

There was no change in Grandmother's attitude after that. When I went near her, I would purposely sigh, my face downcast and so forth, but she never responded.

Having given me such heavy pain, did she never think about it? Or, was she inwardly upset, and deliberately refraining from showing it in her face?

Anyway, the story of Mugiko would end up hurting Shinjirō, too, not just me. That might be why Grandmother pretended that nothing had happened and smothered her feelings, though in fact she wanted to comfort me this way and that—I tried to conjecture in this manner.

Today Grandmother went out again after lunch to visit her elderly friend, purse in hand and towel over her head as usual. Towel over her head—what was happening to Grandmother? It probably played the same role as our hats for shading from the sun, but it was alarming to see her walking through the rice fields, with a towel, folded in half, on top of her thin hair.

Since she was light, she walked hovering insecurely like a leaf. The towel on her head marked Grandmother's location in the rice field.

Naturally she's gone again to that house near the big camphor tree, I thought. She must be chanting, drenched in perspiration, "Gyaatei, gyaatei, sowa sowa gyaatei."

On a whim I tried recalling the names of those close to Grandmother, those who had died early.

Jikurō, the one who was insane;

Tetsurō, who fled through the mountains;

the shoemaker's wife, who had pointed at the cedars of double suicide;

then O-Mugi and her daughter Mugiko;

and Grandfather who had died early.

Many, many others of her siblings that I didn't know would be among them.

It may be that Grandmother was sending up that "gyaatei, gyaatei" to these people. However, I could not believe that weird voices like that, sent to Tetsurō and Mugiko, could console them.

The dead were so strange, I thought.

There were many things I didn't understand.

I visualized the many frogs crying in the field on that "gyaatei, gyaatei" day and the white rose in the yard with the line of ants crawling up along the stalk. I felt that those little things were the ones conducting a real funeral.

The ants climbed the rose stalk steadily, moment by moment. Zigzag, zigzag. . . . At the end of the climb was a rose like a white evening sun.

Even if Grandmother did not chant the sutra, the frogs and ants did.

Nature was like a deep container, I thought.

In the late afternoon, Grandmother came home, swaying unsteadily, with the folded towel on her head and the purse hanging from her arm. About the time she finished washing her sweaty clothes in the bath, a violent evening shower began. Grandmother put her hand to her chest in relief, looking out the window at the outside that had become all dark. If she had been delayed a little, she might by now be drowned in the rice field.

Tateo was playing the organ in the back room. One moment the sound was obscured by the rain, only to be heard again the next minute.

Shinjirō, who had been out, ran home soaked.

The evening shower did not stop for a long time. Both Minako and I heard the thunder in the kitchen and ran to the dining room, leaving the pot and dishes. Grandmother, Shinjirō, Minako, and I found that we were all sitting straight without conscious effort. The thunder sounded as though it was camped right above us in the ceiling.

"Tateo's such a grown-up, I should have known it," Grandmother said, impressed. True, the sound of the defective organ he played continued slenderly amid the sound of the rain and thunder.

At night after the rain was up, there was a call for Grandmother from Tateo's mother, reporting that all three of our respective mothers had returned from Hawaii that day. It seemed that our fathers had returned earlier, since they had to worry about their jobs.

Suzujirō, considerably better, had been switched from intravenous feeding to a liquid diet. He had remembered all sorts of things about Grandmother and told them stories from the old days. Grandmother didn't seem to like that. It must be weird when a stranger knows you well.

"Did she say what Clark was like?" Tateo asked.

"She didn't talk about him," Grandmother said.

"What kind of house do they live in?" Minako asked, leaning forward with interest.

"I didn't hear," Grandmother said.

"Did she say there were children in the family?" Shinjirō asked.

"I don't know," said Grandmother, "If you want to know so much, you should go back home and ask your parents now that they are home."

V

Red fruit was hanging between deep green leaves. Five tomatoes, each like a setting sun, grew in the garden.

"Come, Tami, give me the basket," said Grandmother, and went into the vegetable garden, her back bent. Her upper body was not visible; only her hips in a gray one-piece moved across the garden.

I handed her the basket I held in my hand. This morning Grandmother plucked three tomatoes. These were for me to take home and slice up for breakfast.

"Why don't you pick some farm lotus, too," Grandmother said, so I added five or six of them to the basket. Thin baby hair grew on the roots, which looked like long triangular hats. The hairs pricked any careless fingers that touched them.

"Come, here's an extra," Grandmother threw in the basket a beautifully swollen eggplant. The basket in my hand became that much heavier. Grandmother lifted her bent back, then stretched her body while holding the towel on her head.

This was the signal for us to return.

Leaving the garden, we circled the wall of the house and came to the gate. Although our path was light in the early dawn, insects from the night before were still crying. Since the voices welled up from the grass thicket and rose directly into the sky, they sounded as though falling all over the place from heaven, having been sprinkled generously. As if crossing a clear river, Grandmother and I walked quietly.

When we entered the gate and reached the door, Grandmother finally spoke one simple sentence:

"You don't have to go home?"

"Not yet," I said.

Grandmother's face softened in relief.

In the kitchen Minako was washing some leafy greens.

"All you have to do is cut them up," she said.

At Grandmother's house Minako had developed the habit of rising early. We had decided that we would stay longer after all, and had postponed our departure. I thought that Minako's good habit would prove a nice gift to take home when we did leave.

In passing, although we didn't make much progress in our studies, everyone was able to find something to take home. Shinjirō went out and got tanned, shedding his habit of reading cartoons. This was an earth-shattering change.

Tateo learned to sit on the floor with correct posture when eating—he had grown up with chairs—and to play the organ with both hands.

I learned to manage the meals of five people. I realized that helping was one thing, and managing quite another. Even after I grew up, I thought, this would probably influence the way I looked at and thought about things.

I stood facing the kitchen window.

A cock was crowing from the yard of a distant house.

I felt my heart becoming as clear as the wind. Cutting up the leafy vegetables Minako had washed, I scattered them over the *miso* soup in the pot.

The *miso* soup was unclear as if with mist. It was because the dregs of fermented wheat created a deep fog. Each time I mixed the soup from the bottom of the fog, the dregs rose in swirls. Small pieces of eggplant, fried tofu, and greens bobbed up and down in the muddy current.

As a thought occurred to me, I tried to enlarge in my mind the pot I was looking down at, to an area about the size of the surface of a big pond, and I strained my eyes to gaze into it. I found things in the *miso*-soup flood moving in small flickering motions. From the soup's surface, a yellow ochre as though earth from the farm had been mixed in, a man's head and hands stuck out. They must be Jikurō's head and hands, I thought.

Also surfacing and disappearing was the handle of the metal hammer with which Tetsurō pounded shoes, *tonkachi tonkachi.*

Mugiko's hair was streaming.

Two cedar trees were fast asleep.

The mountains had sunk.

The rice field had sunk.

The house,

the cattle,

and the cock

floated.

They were like dust as I looked down.

I stopped the fire and put the lid on.

Grandmother's pot was frightening.

Tateo, Shinjirō, Minako, and Grandmother all showed up at the table in the dining room. I carried the pot there. "Treated to meals without doing a thing," Grandmother bowed.

Then, she said as if singing,

"Heavenly, heavenly."

When breakfast was over, Shinjirō looked out the window.

"Let's go to the pond behind the temple," he suggested.

We had come here all this way, so we should see everything, he added. Grandmother agreed: "Yes, yes, he's right. Even I can walk there." We agreed that, as long as we were going, we should pack lunch and eat it there. Only Minako seemed a little worried and asked, "Will there be any snakes?" She must have been really terrified by that snake at the fall.

It was fine that morning.

The five of us walked along the path in the rice field with Grandmother in the middle, with a towel on her head. "What happened to that parasol?" Shinjirō asked her. "It's carefully stored away," Grandmother answered.

"Isn't that how men looked in the old days, when they went to public bath houses? It looks funny," Shinjirō said.

"This is the country, so nobody ever looked like this," Grandmother said.

"Why?"

"Because there weren't any bath houses."

"I see. . . ."

"So I won't be mistaken for a man."

We came across someone on the way to the farm.

Grandmother greeted him with a smile. When asked, she looked back at us and said,

"Oh, no, these are just my disobedient grandchildren."

Shinjirō looked unhappy.

"Why did you say such a thing?" he asked.

"You can't praise members of your own family," Grandmother said.

"But people who don't know us may take you seriously."

"They can tell by my face."

"Then, what did you look like when you were talking to that person just now?" Shinjirō asked.

Grandmother twisted her face toward him, and reproducing her expression, said, "Like this—"

Since we walked slowly, keeping pace with Grandmother, Shinjirō gradually became bored. Although he kept looking down at his feet in sandals as he walked along, he finally went ahead by himself.

The pond was behind the wall of the temple where Shinjirō always took a nap. One side of the edge of the pond was unapproachable, covered with thick foliage. The remaining half of the perimeter ran along a small trail, and there seemed to be a spot for resting.

The pond was not so big. It was a size that allowed us, since it was round, to see its entirety. We could also clearly see the birds that came flying to the trees on the opposite side.

With murky dark green water, it was a pond that one probably shouldn't look at when feeling unwell. The sky was so bright, however, that pieces of the clouds reflected on the water floated very white.

"I wonder if it's deep," Tateo said.

"Oh no, it's only up to a man's neck or so," Grandmother said. "It's shallow around the edge. It looks deep, but it doesn't even reach the waist. Then there is a sudden dip, but it's still only up to the chest." She was very informative. "The water gets up to the neck around where those trees hang down. So even if you picked up a drowning child, you couldn't get him out of the water there, but would have to carry him to where it was shallow all the way on this side of the pond."

"Were you watching?"

"Was I watching! He was a student in my class," Grandmother said staring at the pond.

"There were a lot of frogs," she explained, "so children used to come here to catch them. It's awful if you lose your life instead of catching frogs."

There didn't seem to have been any children who died there.

"There used to be lots of children here, so someone was bound to find you."

Here Grandmother's tale turned to the young Jikurō. According to her he had once nearly drowned in this lake.

"I'm going to have some fun, he said as he went out. It was already near dusk, and he came to catch frogs. There were lots of frogs in the shade of the trees over there. He caught a few, and on the way back to this side, he drowned while swimming near the center of the pond. He often got leg cramps."

"There's no hope if you get cramps where the water is neck-deep," Tateo said.

"Did anyone see him?" Shinjirō asked.

"It was so late in the afternoon that there wasn't a single soul at the pond," Grandmother shook her head.

"But something interesting happened," Grandmother continued. "I was in the kitchen. A strange child poked his head through the back door, and called me, hey Si-i-s, Big Si-i-s. In a whispering voice. I turned to the voice, surprised, and found a thin, little boy. Jikurō was drowning in the pond, he said, so I pulled him out. Sure enough, the boy's clothes were soaking wet, water dripping from his hair. He looked at me with an odd expression and smiled. I let him cough the water out and put him down on the grass, so go quickly—"

"He really said that?" Shinjirō asked.

"Yes. Our big brothers ran to the pond right away and found Jikurō, lying drenched on his back on the grass, just as the boy had said."

We looked at one another in wonder.

"How did a child that small pull Jikurō out?" Tateo sounded half in disbelief.

"Later I heard some villagers talking with my father. They decided after a while that it was a kappa, a water imp."

Again we looked at one another.

"Those limbs, lean as cucumbers, and that grassy green face, could belong to no one else but a kappa," Grandmother said.

We fell silent. We had been so engrossed in this tale, but when it ended like that, we couldn't avoid feeling lost. Like finding, while being guided along country roads, a long fox tail behind the trusted guide—it was that kind of odd feeling. "Better be wary of this guide," read the insecure expression that vaguely surfaced on our faces.

"Grandmother, was that boy's face really the color of grass?" Shinjirō asked again, looking doubtful.

"All green," Grandmother rolled her eyes, showing the whites. So Shinjirō joined us in our silence.

A wind started over the surface of the pond, and the sky grew overcast. Dark clouds streamed over our heads. As the old woman's tale was incomprehensible, so the water in the pond, too, had dulled before we realized it.

We opened the lunch packs of rice balls which we had brought. When we were halfway through our lunch, the clouds left again, and the pond lightened up in one sweep. In the middle of our rice balls, Tateo asked Grandmother as if by whim:

"Well, uh, at about what age did this Jikurō person go insane?"

Minako, Shinjirō, and I, following Tateo's example, looked at Grandmother all at once. Right, that was something we needed to ask her about.

For some reason, however, Grandmother lifted her eyes questioningly. She inclined her head.

"Jikurō . . . ?" she asked.

"You remember, you told us how he became insane and was confined to a room like in jail?"

"Such nonsense," Grandmother looked puzzled. "He never became insane or anything. He went all the way through to high school and became a teacher."

We went momentarily blank.

"In the end he became the schoolmaster of a middle school in the northeast, and on his way there he died of cerebral hemorrhage in the train."

"Grandmother, you're lying!" Shinjirō said in a loud voice. "You said he died insane. Remember? He was always writing all sorts of Chinese characters, you said."

Grandmother was still inclining her head.

"Well, I guess I may have told you a story like that."

"Are you kidding? *Eye, mouth, nose,* he wrote, you said."

"But that's probably not Jikurō. I have it in a document from back then that Jikurō collapsed in the train—"

"Then, who was the one who went insane?"

At Shinjirō's loud voice, Grandmother looked startled and stared into a vacuum.

Folding her fingers over her lunch box, she listed once again the names of her thirteen siblings. It was the same as the last time even to the detail of skipping Suzujirō's name. She remembered the names so clearly, yet the important connection between people and names did not match up correctly.

Eye, mouth, nose. . . . Who was writing these characters on paper? What was the name of the boy who was writing *neck, chest, belly* in a corner of a dark, barred tatami room? As it became questionable to what extent we could believe the story of the boy rescued by the kappa, we also became unsure of the story about the boy who wrote *eye, mouth, nose.*

Countless watery creases moved on the surface of the pond. A breeze came toward us. Like these ripples, did everything flow away to a faraway place?

That noon, Tateo and I came home from the pond with a common, though independently formed, question.

Each of us spent the whole afternoon trying to draw an answer from the question. We reached the same conclusion: "Grandmother's memory was a mess."

After supper, I went to the inner room, where Tateo had started to play the organ. I confided in him about the story of Mugiko that Grandmother had told me in the bath. I told him about the melancholy and helpless days that had followed, when I felt as though my feet were hidden by a mist.

Sitting cross-legged and leaning against the organ, Tateo nodded, "I know exactly how you feel, Tami."

He too revealed his days of anxiety after hearing from Grandmother the story of Tetsurō and the love-suicide cedars. If someone like

Tetsurō were his grandfather—from the moment that thought occurred to him, he said, the dry needles of the two cedars had constantly rustled behind his back, whispering to him.

"What kind of thing did they whisper?" I asked.

"I can't tell you." Tateo looked embarrassed. Recalling his offensive behavior when we had gone to see the cedars, I closed my mouth.

"We're up in the air now," Tateo said. "Since we've discovered that her memory's that awful, we can't keep hanging like this. The first question to solve is to what extent her memory is accurate and to what extent it's unreliable," Tateo said resolutely, his usual lazy smile gone. His expression then was to me the most likable of all his faces I had seen.

When we returned to the dining room, Shinjirō and Minako were studying face-to-face for a change.

The incident that day seemed to have made a strong impression on Grandmother. Opening an old album, she was viewing it with a much more serious face than Shinjirō or Minako. She held a magnifying glass in her hand.

Tateo came and sat next to her with his usual expression.

"What are you looking at?" he asked.

Ahh—Grandmother exhaled a long breath and answered, lifting her tired eyes: "I have many people to look for."

"Who are you looking for?"

"First, Suzujirō, then Jikurō," Grandmother said with deep feeling.

"I see," Tateo nodded, looking as though that were a matter of course.

"Then, Mugiko and Tetsurō. . . ."

Ah—now I was the one to want to heave a deep sigh. For Grandmother, when she said this, turned the light of very lonesome, sad eyes toward me.

"Also Sekisuke and Senji."

The light of her eyes deepened into a sadder and sadder color.

It seemed to me like the color of the eyes of a human being who, lost on the way, could not return home.

"Shōsuke and Ginrō, then Ryōtarō . . ." the names continued.

But her words no longer registered in my ear.

I was focusing in my mind on the conclusion I had reached earlier: "Grandmother's memory's a mess." That's right, I thought, isn't her memory in a terrible state? I gazed at her hopelessly as she, with the magnifier in her right hand, started again to fold the fingers in her left hand. As if he had lost the energy to speak, Tateo, too, eyed the motion of her fingers vacantly. There was no longer any way to connect her fragmented memory.

Grandmother's hand caressed the old album. The magnifier kept on enlarging the yellowed faces of lots of people. To me they looked nearly the same. Since they had been born one or two years apart, or in a row at the beginning and end of the same year, it was difficult to recognize their characteristics in the faded photographs.

Once I had wanted to know in what form the memories of the past from decades ago were stored in Grandmother's brain. I had thought, at that time, that something like fog or mist had perhaps filled the space above her memory.

However, now I was able to conjecture that old memories were, like these photographs, something like fragmented film that had lost its proper nouns and was stripped of context.

The night got late. Shinjirō's and Minako's notebooks indicated fair amounts of progress. A cock's night cry streamed into the quiet room. It was an odd cock. It crowed morning, day, and night.

Tateo was lying on his back on the tatami floor, his eyes open. Who was his grandfather after all, I wondered? Was it Tetsurō, who had seen the love-suicide cedars? Or was it all right to leave it as before?

Then my thoughts shifted to myself. Was my mother Mugiko, or my current mother? By now I knew that, no matter how much I questioned my grandmother, I couldn't get an answer that would be sufficiently convincing. Even if we asked our parents, we would never be fully convinced, Tateo said to me. So, he had nothing else to do but stare at the ceiling.

Suddenly I looked by my side at the fair-skinned, neat face of Minako who was racing her pencil across a notebook. I felt that a pretty girl like Minako was more suitable as a daughter to the unfortunate Mugiko. Minako and I were the same age, and both of us were Grandmother's grandchildren. The possibility was the same for Minako, too.

But I felt I would probably never tell anyone about Mugiko. Minako should remain as she was. I, too, wanted to remain as I was. There was no need to clarify forcefully what was not clear. I would live as before, believing my present mother to be my mother.

Arriving at this thought, a big, slow wave of sleepiness finally began to enclose my body.

Shinjirō yawned over his notebook. Minako, as if on cue, carried a hand to her mouth. My hand also moved likewise. In a corner of the room, Grandmother had long been swaying, as if rowing a boat.

She must have rowed out into that pond, I thought. It was the pond of the big, black iron pot. The fog must be thick tonight, like chaotic and terrifying clouds. Grandmother was pushing her boat out there. She was starting to row. Creak, creak, creak—even the sound of the oars could be heard.

Grandmother's boat sped across the surface of the night pond.

Many things were floating there.

Tetsurō's head and hands—no, they might no longer be Tetsurō's.

Let me rephrase: someone's head and hands.

And someone's hair was flowing.

A metal hammer with which someone pounded shoes was bobbing.

The double suicide cedars that somebody had seen were fast asleep.

Making the mountains sink, making the rice fields sink, and making the house, cattle, chicken, and dust sink, Grandmother would row the boat.

There was no way anyone could match her. . . .

It had been hazy since morning, with mosquitoes whining around. In the afternoon a reply to Grandmother's letter came from Clark.

Opening the letter, Tateo started to turn his head in puzzlement.

The contents were as follows:

Dear Mrs. Hanayama Sanae:

Thank you for your letter both my father and I read it. But we don't understand the characters and we don't understand the sentences.

My father was very happy thank you. His illness will get better quickly please write. Everyone will be happy and read. Well now by way of reply. Good-bye.

Clark

This would get us nowhere.

We decided that Tateo would rewrite Grandmother's letter. Bringing to the verandah mosquito coils, stationery, and a fountain pen, Tateo lay on his stomach. It was difficult to compose a letter for someone else. For half an hour, Tateo only stroked the tip of his nose with his fountain pen while he lay staring at the stationery, but eventually he corrected his grip on the pen.

And he finished the letter in five minutes.

His version read as follows:

Dear Clark:

Thank you for your letter. Everybody was happy to read it. I am fine, too. I hope Suzujirō will get better soon. I want to see my younger brother's face. Please send a couple of pictures. I shall be waiting. Good-bye.

Hanayama Sanae

"You call this ghostwriting?" I rounded my eyes.

"It's meaningless if they can't read it," Tateo responded.

"You pondered half an hour. What a waste."

"Let's seal it quickly before Grandmother demands to see it," Tateo said.

Shinjirō came running. "Let me do it," he said, snatching the envelope.

He stuck out his tongue and licked the gluey flap.

"I like this kind of thing," he said, then sealed the envelope by pounding it with his hand.

Supper was going to be a combination dish of kelp, Koya-style tofu, and beef, plus cucumber salad. Toward the evening I went to the garden in back to pick cucumbers.

The atmosphere was heavy, and it felt as though I were inhaling thick air. Even as I was picking the cucumbers, mosquitoes started an assault. I waved my arms and stamped my feet as I picked. Under the dense air, the cucumbers looked deeper green than usual. The clouds in the sky looked as though they would release drops if squeezed.

As I came out of the patch, a heavy evening shower started.

I started to run pressing the basket with cucumbers to my chest. The sky darkened gradually starting from the direction of the distant rice field, and the rain came pressing toward me. I clearly saw the rain approach like a bamboo curtain, swallowing rice paddy after paddy. It resembled the way human legs run.

When I turned the path, I spotted something in the distant dark rice fields: something white fluttering in the fields almost as dark as night. I stopped. Straining my eyes, I saw a human body under the white thing. Grandmother, I thought. She was running home with a white cotton cloth on her head.

The white flutter, however, came no nearer. No matter how her legs ran, they never got faster. Only her body was dancing. It must be tiring.

The rice paddies, viewed from afar, were already flooded. They formed an ochre-colored pond. In their midst, Grandmother's white cloth kept dancing. For a while I looked at that white color. Then, placing the cucumber basket in front of the gate, I took off the kitchen clogs I was wearing.

Then I started running to help Grandmother out.

Glossary

ayu	river fish, which swim upstream toward the rapids in spring
bijoh	military term for belt buckle, obsolete
chokori	Korean women's blouse
furisode	long-sleeved kimono normally worn by unmarried young women
furoshiki	square cloth used for wrapping
haori	lined jacket worn over a kimono
hiryōzu	also *hiryūzu,* fried dumpling made of rice powder or flour mixed with tofu or shredded yam, or simply fried tofu; from Portuguese *filhos*
imagawa-yaki	small drum-shaped hotcakes with sweet bean inside
janome	"snake's eye," two circles with a dot in the center, a design often used for umbrellas
jō	death through Zen meditation
kaimaki	kimono-shaped comforter with sleeves
kappa	legendary creature living in a river; usually impish
karatachi	"Chinese orange," thorny, trifoliate bush with small, five-petaled white blossoms in late spring and non-edible yellow fruit in autumn

kasuri	cloth woven or dyed with splashed white patterns against deep color background
keyaki	"zelkova," tall deciduous tree, with pointed oval leaves with zigzags; grows naturally in Honshu, Kyushu, and Shikoku
kichanamashii	dialectal expression meaning "impure"
komageta	a pair of clogs each carved out of a single block of wood
kotatsu	heater covered with a comforter for people to sit around
kudzu	wild perennial vine that blooms in early autumn, or, as here, starch taken from its root resembling arrowroot
meisen	a kind of thick unfigured silk often of striped or splashed pattern
miso	soybean paste
monogatari	tale
nattō	fermented soybeans usually eaten with hot rice
obi	broad belt worn over a kimono
obikawa	leather belt
ohagi	sweets made of steamed sticky rice covered with red bean paste or powdered toasted soybeans, associated with the equinox festivals
ojisan	"uncle," a friendly address for an adult male
okusan	a common address for a housewife
senmu	executive director
setta	leather-soled sandals woven of bamboo skin with metal reinforcement at the heels

shamisen	three-stringed traditional instrument
shikishi	square poetry card
shōji	sliding screen doors mounted with one layer of white paper
susa	hemp or straw used as an ingredient in strengthening mud walls
tabi	divided socks
taikaku	military term for leather belt, obsolete
takashimada	a traditional hairdo with a high bun in back, worn by formally dressed young women
tanabata	July 7 festival
tokonoma	alcove, or raised floor sequestered in a corner of a room for hanging a scroll and placing arranged flowers
tsukudani	preserved small pieces of seafood and other food prepared with soy sauce
umeboshi	pickled plums
yukata	"bathrobe," or one-layered cotton summer kimono worn in the evening

About the Authors

Enchi Fumiko (1905–1986)

Born in 1905 in Asakusa, Tokyo, the daughter of a famous scholar of Japanese literature. Withdrawing from high school at the age of seventeen, finding that school taught her little, she studied English, French, and Classical Chinese with a tutor. In 1926 her play *Furusato* (Native Land) was selected for publication by the journal *Kabuki*. Joining the seminar in playwriting given by Osanai Kaoru, a leader in the Modern Theater movement, she continued to write plays. In 1927 she married Enchi Yoshimatsu, and having joined the little magazine *Nichireki,* started to write fiction. Her first book of short stories, *Kaze no gotoki kotoba* (Words Like the Wind) was published in 1939.

After that, Enchi entered a long period of silence which did not end until 1953, when *Himojii hibi* (Hungry Days) was published, receiving the Women's Literature Prize. In the works that followed, *Yō* (Eeriness, 1957), *Onna zaka* (Waiting Years, 1957, the Noma Hiroshi prize), *Nise no en shūi* (Love in Two Lives: The Remnant, *Bungakukai,* 1958), *Namamiko monogatari* (1959), *Hanachiru Sato* (1961, the Women Writers' Association prize), and the trilogy *Ake o ubau mono* (The Thing Which Removes Red, 1955–56), she delved deeply into the psychic realm of women, developing her own fictional space spanning the gap between classical and modern Japanese literature. In 1967 she began translating *The Tale of Genji* into modern Japanese language, a work which she completed in 1972.

In 1972 Enchi received the Grand Prize of Japanese Literature and continued her prolific creative life, despite a series of illnesses which she combated with determination. In 1970 she was made a member of the Academy of Art.

Her last works included *Shokutaku no nai ie* (A House without a Dining Table, 1979), *Yuki moe* (Burning Snow, 1977), and *Karasu gidan* (Casual Talks by a Crow, 1981). She was a fine essayist and literary critic, especially of classical Japanese literature.

Hayashi Fumiko (1903–1951)

Born the daughter of a peddler from Ehime Prefecture, who later ran a store in Wakamatsu City. Her mother, Kiku, whose fourth marriage it was, ran off with her husband's store manager, twenty years her junior, taking her daughter Fumiko, then about ten years old.

The family plied its trade in various places, and Fumiko attended several different schools, taking eight years to finish the six-year elementary education.

Hayashi Fumiko started to write poetry for local newspapers while in high school. After graduation she went to Tokyo to join a Meiji University student whom she had known in Onomichi. On graduation, however, he broke off their engagement. She worked as a maid, factory worker, saleswoman, barmaid, and cafe waitress while continuing to write poetry. The diary she began around the time her lover left became the basis of her first novel, *Hōrōki* (Record of Wandering, 1928–30).

In 1924 she became acquainted with a group of anarchist poets, some of whom became her lovers. The group, which met regularly in a restaurant above the Nantendō bookstore in Hongō, Tokyo, included Tsuboi Shigeji, Takahashi Shinkichi, Hagiwara Kyōjirō, Tanabe Wakao, and Tomotani Shizue. With Tomotani Shizue, she published a poetry leaflet, *Futari* (Twosome), which lasted three issues. Her poem "Jokō no utaeru" (A Factory Woman Sings) also appeared in *Bungei sensen* (Literary Front), and she came to number leftist writers among her close friends.

She married Tezuka Ryokubin, a painter, in 1926 and in the following year for the first time settled in a home. Not only did she remain a roamer in spirit, but also made many long trips within Japan and to Europe, China, and Southeast Asia, including trips as a reporter at the front during 1937–43. With Sata Ineko and other authors she participated in the Asahi Shinbun trip to Manchuria to comfort soldiers in 1941 and in another similar trip to Southeast Asia in 1942–43.

The first of her two poetry anthologies, *Aouma o mitari* (I Saw a Pale Horse), was published in 1929. In October 1928 the first portion of her autobiographical novel, *Hōrōki,* appeared in *Nyonin gei jutsu,* a journal for women writers. While "Fūkin to sakana no machi" and "Seihin no sho" (both 1931) are autobiographical extensions of *Hōrōki,* other outstanding short stories are pure fiction finished with artistic mastery: "Kaki" (Oyster, 1935); "Bangiku" (Late Chrysanthemums, 1948—Women Authors Prize, 1949), which depicts a fifty-seven-year-old woman's disillusioning reunion with her former lover; "Aibiki" (Tryst, 1946); "Kawahaze" (River Smelts, 1947); "Shitamachi" (Downtown,

1949); "Hone" (Bones, 1949, tr. 1966); and "Suisen" (Narcissus, 1949) number among her most highly regarded short stories.

She maintained the popularity she earned with *Hōrōki* throughout her life by repeatedly capturing, with compassion, the poverty, darkness, despair, and misery of war, especially of women moving from place to place, from job to job, from man to man. She died at age forty-eight before completing *Meshi* (Meals), which was being carried in installments in the *Asahi shinbun*.

Hayashi Kyōko (1930–)

Hayashi Kyōko was born in Nagasaki in 1930. She spent her prewar and wartime childhood in Shanghai. Returning to Nagasaki in March 1945, five months before the war ended, she attended Nagasaki Girls High School. Hayashi was working at a munitions plant in Nagasaki when the atomic bomb was dropped on August 9. She was seriously ill for two months, and like the majority of bomb survivors, suffered thereafter from fragile health and the threat of the aftereffects that frequently produced birth defects in offspring. She started to write in 1962.

She made her literary debut with "Matsuri no ba" (Ritual of Death) in the June 1975 issue of *Gunzō*. It received the *Gunzō* magazine's Prize for New Authors and subsequently, the Akutagawa Prize. It records her exodus from the area of instant devastation and her reunion with her family.

Hayashi continued to write of the atomic bomb in a short story "Nanjamonja no men" (Mask of Whatchamacallit, *Gunzō*, February 1976), and a sequence of twelve short stories carried in *Gunzō* from March 1977 to February 1978 and published in book form under the title *Giyaman bīdoro* (Cut Glass, Blown Glass) in May 1978. The fifth story, "Kōsa" (Yellow Sand), dealing with the author's experience in Shanghai, has a special place in the sequence. The other eleven stories concern the bombing, with the only reference to her Chinese experience found in a flashback in "Hibiki" (Echo).

Although in most of her works Hayashi has concentrated on recording the effects of the bombing, she views the Chinese experience as something that also contributed to the shaping of her mind. What she saw as a schoolgirl there, too, was part of her war experience as a whole. *Missheru no kuchibeni* (Michelle Lipstick, 1980) is a fuller account of her girlhood in Shanghai. *Shanghai* (1985 winner of the Women's Literature Prize) is a travelogue based on a five-day trip to the city thirty-six years after she had last seen it.

In *Naki ga gotoki* (As If Nothing Had Happened, 1981), a full-length novel, Hayashi refers to her determination to be Nagasaki's "chronicler." Her effort as a chronicler in the sixties and seventies was mostly directed toward recounting the bombing, survivor's psychology, and the fate of her friends and teachers. In the eighties her topics become more diverse: marriage, birth, divorce, her grown son's marriage, birth of their child, the environment, aging, and death, which are all intricately connected in her mind to the bombing and war. Some stories in *Michi* (The Road, 1985) are based on her thoughts on her son who is an "A-bomb nisei," and her divorced husband. *Sangai no ie* (No Abode, 1985), whose title story won the 1983 Kawabata Prize, adds depth to Hayashi's study of human psychology through the theme of her father's death and the relationship between her parents. Stories in *Tanima* (The Valley, 1988) and *Vajinia no aoi sora* (Blue Sky of Virginia, 1988) reflect her experiences during her three years of residence near Washington, D.C. *Yasurakani ima wa nemuritamae* (Rest Now in Peace, 1990) is a requiem for a teacher whose journal recording mobilized girl students' factory work was found thirty years after her death, and a farewell to those eleven weeks that ended in the bombing.

Hirabayashi Taiko (1905–1972)

Born in Sunawa, Nagano, to a farmer's family. From the time of her elementary school days, she read widely and was determined to be a writer. By the age of nineteen she was exposed to socialist writings in *Tanemakuhito* and planned to quit school to join the movement. At seventeen (in 1922) she moved to Tokyo and while working as a telephone operator studied English at night school.

Several anarchists became her friends, and one of them became her lover. She was arrested several times. In 1924 she moved to Manchuria where she had a child who died in infancy. After returning to Tokyo she lived with another anarchist and started writing for *Bungei sensen* (Literary Front) in 1925. In 1926 she joined the Proletarian Art Association, and the next year married Kobori Jinji, an activist in the proletarian movement. Later that year, her story "Seryōshitsu nite" (At the Clinic) received a Literary Association prize, giving her recognition as a promising proletarian writer.

Hirabayashi continued to write proletarian literature prolifically until 1938, although she had moved away from the *Bungei sensen* because of its internal conflicts. In 1937 and 1938 she was imprisoned for eight months, during which time she contracted tuberculosis. As a consequence she was unable to continue her writing until 1946. "Blind Chinese Soldiers," published in the March issue of *Sekai bunka* (World Culture) in 1946, was one of the first stories she wrote after recovering from her illness.

In 1945 Hirabayashi joined the newly formed left-wing writers' association, Shinnihon Bungakukai, but disappointed with its views on literature, she left the group. After that, she turned to autobiographical works and wrote much on women's experiences. In 1955 she was divorced and in the same year she became involved in the movement to abolish legal prostitution. In 1968 she received the Women's Literature Prize (for "Himitsu" [Secret]) for the second time.

In her later years she traveled to various Asian countries to give lectures. Her major works include *Kō iu onna* (Such Women, 1946), for which she received the Women's Literature Prize for the first time, *Kanashiki aijō* (Sad Love, 1963), and *Tetsu no nageki* (The Lament of Iron, 1969).

Kōno Taeko (1926–)

Born in 1926 in Osaka into a wholesale merchant's family. From her infancy she was physically weak and subsequently also had difficulty adjusting psychologically to school life. During the war she was drafted as a student worker, and she spent her adolescent years working at a military-related clothing factory. Having failed in the entrance examination for the Japanese literature department of Osaka Women's College, she entered its economics department in 1940.

In the postwar period she experienced a strong sense of failure and dissatisfaction. In 1950 she joined the literary magazine *Bungakusha* (Writers), which had just started in Tokyo. She found difficulty in writing consistently until 1961, however, since she had contracted tuberculosis and because of the demands of her job at a government office. In 1961 "Yōjigari" (Hunting for Infants) was published in *Shinchō,* and in February 1963 "Bishōjo" (Beautiful Girl) was nominated for the Akutagawa prize. In August of the same year she received the Akutagawa Prize for "Kani" (Crabs). "Ari takaru" (Ants Swarm) was published in the May issue of *Bungakukai* in 1964. In 1967, she received the Women's Literature Prize for *Saigo no toki* (The Last Moment), a collection of several short stories which included "Ants Swarm."

Kōno has continued writing actively up to the present as a major novelist, literary critic, and playwright. In many of her works she delves into existential questions on the anxiety of life, which she develops into pathological dramas of women who are unable to conceive.

Her major works include *Saigo no toki* (The Last Moment, 1966), *Kaiten tobira* (The Revolving Door, 1970), *Hone no niku* (The Flesh of the Bone, 1971), *Mukankei* (No Relations, 1972), *Chi to kaigara* (Blood and Shells, 1975), and *Miira Tori: Ryōkitan* (Mummy Hunting: A Story of the Grotesque). In 1976 she received the Yomiuri Literature Prize for her essay "Tanizaki bungaku to kōtei no yokubō" (Tanizaki's Literature and the Desire for Approval), and in 1980 she received the Tanizaki Prize for her novel *Ichinen no bokka* (One Year's Pastoral Song).

Miyamoto Yuriko (1899–1951)

The first child of a noted Cambridge-trained architect, and granddaughter of Nishimura Shigeki, a scholar of ethics, a leading intellectual of the Meiji period, and a cofounder of *Meiroku zasshi.* At the age of twelve, the precocious Yuriko started writing short stories. Her first novel, *Mazushiki hitobito no mure* (A Flock of Poor People), written when she was seventeen years old and a freshman at Japan Women's College, was published in *Chūō kōron* (January 1916) with a strong endorsement by Tsubouchi Shōyō, instantly making her a famous writer. The novel, which is based on her own experiences at her grandfather's estate in Fukushima Prefecture, depicts the misery of the poor peasants and the hypocrisy of the dominant class.

In 1918 Miyamoto accompanied her father to New York, where she married Araki Shigeru, a scholar of ancient Persian language, despite her parents' objection. Their marriage brought five years of psychological struggle and creative stagnation which continued until her divorce in 1924. Her experience of marriage and divorce provided the material for her first major autobiographical work, *Nobuko* (1926).

In 1927 Miyamoto left for Soviet Russia accompanied by Yuasa Yoshiko, a translator of Russian literature with whom she had been living since her divorce. On returning to Japan in 1930, she joined the All-Japan Proletarian Artists' Association (NAPF). In the same year, she became the coordinator of the Women's Committee of NAPF and the editor of the journal *Hataraku fujin* (Working Women). Her life after the divorce and her complex relationship with her woman friend are dealt with in *Futatsu no niwa* (The Two Gardens, 1947), an autobiographical sequel to *Nobuko,* while her experiences in Russia form the basis of *Dōhyō* (Road Sign, 1950).

In 1930 she joined the Japanese Communist Party and in 1932 married Miyamoto Kenji (1908–), a young communist and literary critic. From 1932 on Yuriko's works became the subject of strict control, and she was arrested repeatedly between 1932 and 1942. Kenji was arrested in December 1933 and remained imprisoned until 1945. Despite these hardships and the torture she experienced in prison, Miyamoto refused to give up her ideological beliefs.

During the war she devoted herself to literary criticism and essays on women and women writers, the latter of which were collected in 1948 under the title *Fujin to bungaku* (Women and Literature). The voluminous letters Yuriko and Kenji exchanged during these years were collected in *Jyūninen no tegami* (Letters of Twelve Years, 1951).

In 1942 she was again arrested, but four months later she was sent home in a condition of unconsciousness caused by heat stroke and suffocation; she lost her eyesight for an entire year and her heart was damaged. The experiences of these years and of the confused period at the end of the war were the basic materials for *Banshū heiya* (The Banshū Plain, 1947), while *Fūchisō* (1947) treats her reunion with Kenji.

The years between 1945 and her sudden death in 1951 at the age of fifty-one were the most active and productive years of her life. She helped to establish *Shin Nihon bungaku kai* (New Japanese Literature Association), a group of writers who sustained their anti-imperialist struggle, and took part in the democratic movement as an original founder of *Fujin minshu kurabu* (Women's Democratic Club) and the editor-in-chief of *Hataraku fujin* (Working Women). Most of her major novels, including *Futatsu no niwa, Banshū heiya, Fūchisō,* and *Dōhyō,* all of which trace her life since *Nobuko,* were written during this period. In 1947 she received the Mainichi Shuppan Bunka Shō (The Mainichi Publishing Culture Prize) for *Fūchisō* and *Banshū heiya.*

Miyamoto's short stories include such socialist-realist works as "Koiwai no ikka" (The Family of Koiwai), "Hiroba" (The Plaza), "Sangatsu no daiyon nichiyōbi" (The Fourth Sunday in March), "Yoru no wakaba" (Young Leaves at Night), and "Chibusa" (Breasts).

Yet her major achievement is clearly in her autobiographical novels, throughout which the protagonist tries to liberate herself from her own feudal-bourgeois class background and to contribute to human welfare by fighting against sexism, war, and exploitation of the working class.

Murata Kiyoko (1945–)

Murata Kiyoko was born in Kyushu in 1945. After graduating from Hanao Middle School, she supported her grandmother by delivering newspapers and working as a movie usher and a waitress in a coffee shop. She received the 1976 Kyushu Art Festival fiction prize for her story "Suichū no koe" (Voice under Water). In 1985 she started her personal periodical *Happyō* (Publishing). "In the Pot," winner of a 1987 Akutagawa Prize, was published in book form under that title in the summer of the same year, along with "Suichū no koe," "Netsuai" (Passion, 1985), and "Meiyū" (Sworn Friends, 1987). *Shiroi yama* (White Mountain, 1990) collects seven stories written between 1982 and 1990.

Murata has created many distinctive characters and situations: a brother and sister fascinated by the shape of a cable car; a hilltop world where a little girl suddenly vanishes; two schoolboys engrossed in cleaning school toilets; a boy whose passion is a motorbike; a woman who, after losing her little daughter in an accident, finds a purpose in life in protecting other people's children, and becomes so obsessed and aggressive that she ends up the object of a witch hunt in the housing development where she lives. And more.

Murata's treatment of adolescence is especially notable. Instead of writing memoirs of youth from an adult's point of view, her adolescents tell their stories as they experience growing up. First-person stories like "In the Pot," "Netsuai," "Meiyū," and "Kōsaku densha" (Cable Car) show her gift for presenting a convincing world of juveniles, whether the narrator is a boy or a girl.

Murata also has a special interest in portraying old women. Besides the humorously yet poignantly characterized grandmother of "In the Pot," we encounter ten old women in the title story of *White Mountain.* In this story, a married woman with a daughter is invariably drawn to old women, even haunted by them. One of them is the narrator's dead grandmother, whom she associates with mountains she has climbed or has been attempting to drive up. On the plateau of a lime mountain covered with silvery white pampas grass she indulges in an illusion that she is walking over the white-haired scalp of her grandmother. In the earlier story grandmother's blurred memory was likened to a pot of soup, then to a pond. In "White Mountain," the grandmother is almost the earth itself.

Nogami Yaeko (1885–1985)

Born in 1885 in Usuki (Kita-ama), Ōita Prefecture, the daughter of a prosperous wine maker. While attending the four-year higher elementary school, she studied Chinese and Japanese classics privately with Kubo Kaizō (Chihiro) and composed *tanka* as the only child member of Kubo's poetry circle, Mizuho-kai. In 1900 Yaeko went to Tokyo to join her uncle, an economist with an American Ph.D., and through him met novelist Kinoshita Naoe, who persuaded her to enter Meiji Jogakkō, a Christian-oriented girls' school which fostered many literary figures of the era. While at this school, Yaeko came to know Nogami Toyoichirō (Kyūsen, 1883–1950), a scholar of English literature and Noh drama who was a disciple of novelist Natsume Sōseki. When Yaeko married Toyoichirō, still a Tokyo University student, she had the opportunity to show her novice pieces to Sōseki, who helped publish some of them in

Hototogisu, a literary magazine edited by haiku poet Takahama Kyoshi.

In the 1910s Nogami was already writing for such influential magazines as *Chūō kōron* and *Shinchō,* as well as the new feminist magazine *Seitō* (Bluestockings). Her dramatic works include modern plays exploring contemporary social problems like "Hōka hannin" (The Arsonist, 1916), and others based on Noh drama like "Kantan" (1920).

Nogami shocked her readers with *Kaijinmaru* (Neptune, 1922), a novel based on a real event involving the four-man crew of a fishing boat in which she explored the dilemma of starvation or survival by eating human flesh. *Machiko* (1928–30) concerns a sociology student who, rejecting the middle-class values of her family, runs off to marry a peasant activist. The novel ends, however, with Machiko nearly embracing the class she once rejected in reaction to her revolutionary lover's relationship with another woman. The novel is often contrasted with Miyamoto Yuriko's *Nobuko* (1924–26), which more positively narrates a young woman's self-liberation from traditional values.

The question of morality in the activist movement continued in Nogami's major work *Meiro* (Maze, 1948–56, for which she received the Yomiuri Literary Prize, 1957), in the guise of *tenkō* (political conversion). The relationship between artist and patron is the main theme in her outstanding historical novel *Hideyoshi to Rikyū* (Hideyoshi and Rikyū, 1962–63, Women's Literature Prize, 1964), which she completed at age seventy-eight.

Other important works include "Kanashiki shōnen" (Sad Boy, 1935), "Meigetsu" (Full Moon, 1942), "Kitsune" (Foxes, 1946, tr. 1957), "Fue" (Flute, 1964), "Suzuran" (Lilies of the Valley, 1966), and *Mori* (The Woods, 1985). A nonactivist intellectual who lacked such dramatic personal experiences as those that enriched the lives of some contemporary women writers, Nogami never became a sensation in literary circles. Rather, she stands out for her seventy years of steady, controlled authorship and constant exploration of new horizons. Her last essay appeared in *Chūō kōron*'s January 1985 issue. Her collected works (1980–89, Iwanami) consist of twenty-three volumes of fiction and essays, seventeen volumes of diary, and nine volumes of translation and juvenile stories.

Ohba Minako (1930–)

Born in 1930 in Tokyo, the first daughter of a doctor. Because the family moved from place to place, she had to change schools many times. She was conscripted in 1945 to aid victims of the Hiroshima atomic bomb. Ohba started writing poetry while a student at Tsuda College, but then she married and accompanied her husband to his job in Alaska in 1959, where she stayed except for brief periods until 1970. Although she studied art at the University of Washington at Seattle during these years, she spent her life basically as a housewife, frustrated in her search for self-expression. In 1968, her work "The Three Crabs" received the Gunzō New Writer Prize and subsequently the Akutagawa Prize, marking her brilliant literary debut.

In *Urashimasō* (1977), one of her major works, she pursues the theme of the search for one's personal and cultural roots from the perspective of those whose superficial national identity became blurred. Basically a poet, she explores the fundamental question of the relation between the self and the world, ambitiously attempting to place her search in the archetypal world of human tradition.

Her major works include *Funakui mushi* (Ship-Eating Termites, 1969), *Sabita kotoba* (The Rusted Words, poetry, 1971), *Shikai no ringo* (The Apple in the Dead Sea, Plays, 1973), and *Kiri no tabi* (The Foggy Journey, 1981), for which she received the Women's Literature Prize. "The Smile of a Mountain Witch" was published in the January issue of *Shinchō* in 1976.

Ōta Yōko (1903–1963)

Ōta Yōko was born as Hatsuko in rural Yamagata, Hiroshima Prefecture, by her mother Yokoyama Tomi's second marriage to Fukuda Takijirō, a middle landowner. When her parents divorced in 1910, her mother returned with her to the Yokoyamas, also of Yamagata, and she was adopted by the Ōtas of the same area. Later, she went to live with her mother, who had remarried again. Ōta's stepfather, Inai Hōju, was a well-read, prosperous landowner in Kushima village, Saeki, Hiroshima.

Ōta developed an interest in literature early in her life, and spent many hours of her elementary school days reading in her stepfather's storeroom. After graduating from a girls' school at the age of eighteen, she taught sewing at an elementary school and took various secretarial jobs, moving frequently among Tokyo, Osaka, and Hiroshima. In 1925 she married a dentist, whom she soon learned was already married with three children. She left him and put her first and only child up for adoption. Ōta started to write serious fiction around 1929, when a short story "Seibo no iru tasogare" (Virgin Mary in Twilight) was published in *Nyonin Geijutsu.* She moved to Tokyo in 1930 and continued to write for this and other magazines. Following the breakup of her second marriage in 1937, she began writing her first important novel, the semi-autobiographical *Ruri no kishi* (Shore of a Wandering Journey, published in book form in 1939; made into a radio drama, then a movie, both in 1956). Athough she had been writing professionally for nearly ten years, in 1937–38, still struggling for direction and recognition, she entered two amateur literary competitions under a pseudonym, and won both. Her most memorable works, however, date from the postwar years.

In January 1945 Ōta left Tokyo, a target of air raids, to join her younger sister in Hakushima, Hiroshima City, and there experienced the atomic bomb. From that point on she concentrated on writing about the atomic bomb. In the fall of 1945 she wrote *Shikabane no machi* (City of Corpses; translated by Richard Minear and included in his *Hiroshima: Three Witnesses*, 1990). The novel was censored and only published three years later with portions deleted. Its complete version did not appear until 1950. This was followed by *Ningen ranru* (Human Tatters, 1951, Women's Literature Prize) and *Yūnagi no machi to hito* (City of the Evening Calm and Its People, 1954). "Han-ningen" (Half Human, 1954, Peace Culture Award), based on her own hospitalization, portrays the struggle with mental illness of an author threatened by radiation disease and fears of impending world war. "Zanshū tenten" (Residues of Squalor), first published in the March 1954 issue of *Gunzō*, appeared in *Han- ningen* along with the title story that year. The four-volume *Ōta Yōko shū* (Works of Ōta Yōko), edited by Sata Ineko et al., was published in 1981.

Sata Ineko (1904–)

Born in Nagasaki City in 1904 as the first child of the fourteen-year-old daughter of a Saga postmaster and the eighteen-year-old son of the Miike Mines Hospital director. The following year her mother died of tuberculosis. Between 1916 and 1924 she took various jobs in Tokyo and in Hyōgo Prefecture in a noodle shop, a restaurant, a knitting mill, and the Maruzen Bookstore. In 1922 her poems were published for the first time in *Shi to jinsei* (Poetry and Life).

Sata was briefly married between 1924 and 1925 to a son of a wealthy family, then a student of Keiō University, but her marriage ended after their attempted suicide. In 1926 she became acquainted with members of *Roba* (Donkey, 1926–28), an important journal in the proletarian literature movement, and she married one of them, Kubokawa Tsurujirō (1903–1974). In 1928 her first short story "Kyarameru kōba kara" (From the Caramel Factory, 1928), was published in *Puroretaria geijutsu* (Proletarian Art).

In the early thirties Ineko was an active participant in the proletarian literary movement. With Miyamoto Yuriko, she worked with the women's committee of the Proletarian Authors' Association (1929–31), and she also edited the proletarian women's bulletin *Hataraku fujin* (Working Women, 1931–32), while maintaining contact with underground activists, including Kobayashi Takiji and Miyamoto Kenji (1932–33).

Her middle-length novel *Kurenai* (Scarlet, 1936) portrays her life during the days of withering pressure, conversion, and marital crisis ending in divorce in 1945. She carried the theme into the postwar years with *Haguruma* (Cogwheel, 1958–59) and *Haiiro no gogo* (Gray Afternoon, 1959–60).

In 1940–43 following her arrest and conversion, Ineko traveled in Korea, China, and Southeast Asia. Some of these trips, in which she participated with Hayashi Fumiko and other authors, were organized by the military authorities for the purpose of comforting soldiers at the front. For this she was criticized after the war, and when the predominantly leftist New Japanese Literary Association was founded in 1945 by Nakano Shigeharu, Miyamoto Yuriko, and others, she declined to join it as a gesture of self-criticism.

In 1946 at age forty-two Sata reentered the Communist Party and resumed writing, but was expelled in 1951. During this postwar period she wrote candidly on the theme of wartime collaboration and her disillusionment with the party, for example in "Yoru no kioku" (Memory of a Night, 1955).

Sata Ineko was awarded the Joryū Bungaku Shō (Women's Literature Prize) for her collected short stories, *Onna no yado* (Woman's Abode, 1963), the Noma Hiroshi literary prize for *Juei* (Tree Shade, 1970–73) which traces the impact of the Nagasaki atomic bomb in the minds of two individuals, and the Kawabata Yasunari prize for *Toki ni tatsu* (Standing in Time, 1975). Her other important works include *Keiryū* (The Ravine, 1964), *Omoi nagare ni* (In a Heavy Current, 1970), *Kinō no niji* (Yesterday's Rainbow, essays, 1978), and *Omou-dochi* (Friends, 1989). Her complete works in eighteen volumes appeared in 1977–79 (Kōdansha).

Takahashi Takako (1932–)

Born in 1932, the first daughter in an architect's family. She majored in French literature at Kyoto University and on graduation married the novelist Takahashi Kazumi in 1954. She received an M.A. in French literature in 1958 with a thesis on Mauriac. While Takahashi Kazumi, who received the Bungei Prize in 1962 at the age of thirty, had started an active writer's life, Takahashi Takako continued to work mainly on essays and translations until 1969, when her short story "Kodomo sama" (Honorable Child) was published in *Gunzō.* After Kazumi's death in 1971, however, she plunged into a prolific creative life, writing novels, short stories, and essays. In 1973 she received the Tamura Toshiko prize for her novel *Sora no hate made* (To the End of the Sky). "Sōji kei" (Congruent Figures) was published in 1971 in the May issue of *Bungakukai.*

In 1975 Takahashi became a Catholic, and she is currently living in a convent in France.

Her world is highly fantastic and psychological, with a pursuit of evil which is often developed into frightening dramas of lonely women indulging in a dreamworld to escape from the sterile reality of life.

Her major works include *Kanata no mizuoto* (The Faraway Sound of Water, 1971), *Ningyō ai* (Love of Dolls, 1976), *Yūwaku sha* (Temptor, 1976) for which she received the Izumi Kyōka prize, *Ronrī uman* (Lonely Women, 1977), for which she received the Women's Literature Prize, and *Ten no umi* (The Lake of Heaven, 1977), and *Ikari no ko* (A Child of Anger, 1985).

Tomioka Taeko (1935–)

Born in Osaka in 1935 into a scrap iron dealer's family in a working-class neighborhood. While attending Osaka Women's College as an English major, she showed her poems to the poet Ono Tōzaburō. With his encouragement, in 1957 she privately published her first anthology, *Henrei* (Reciprocal Courtesy), for which she received the H poetry prize. On graduating, she taught high school briefly before moving to Tokyo in 1960. She quit writing poems soon after her 1964–66 visit to the United States during which she lived in downtown Manhattan for ten months. Her complete poems were published in 1967. Tomioka's first fictional work, "Facing the Hills They Stand," appeared in 1971. With this middle-length story of two generations of a family that settled by an Osaka river, she distinguished herself as a superb storyteller and stylist. She has continued to produce numerous prose works, including novels, short stories, movie and radio scripts, plays, essays, and an autobiography.

In a 1976 essay, Tomioka states that she wrote in order to exonerate herself from the fragmented human landscapes that had accumulated in her mind and burdened her since childhood. Surrounded by people to whom these fragments meant nothing, she felt the urge to put them into words. At times some of her characters embody life's incomprehensible situations; others represent a fear of, as well as attraction to, the unknown, the alien, and the insane in daily life, a trace of which they may recognize in themselves, finding it all the more threatening.

Tomioka's prose works include two full-length novels, *Shokubutsusai* (Ritual of Plants, 1973) and *Kochūan ibun* (Strange Accounts of Kochūan, 1974); a semiautobiographical tetralogy, *Meido no kazoku* (A Family in Hell, 1974), for which she received the Women's Literature Prize; and collections of short stories, *Oka ni mukatte hito wa narabu* (Facing the Hills They Stand, 1976), *Shikake no aru seibutsu* (Still Life with a Device, 1973), *Dōbutsu no sōrei* (The Funeral of an Animal, 1975), *Tōsei bonjin den* (Histories of Common People Now, 1977, Kawabata Yasunari Prize), *Hanmyō* (Blister Beetle, 1979), *Sūku* (Straw Dog, 1980), *Tōi Sora* (Distant Sky, 1982), *Nami utsu tochi* (Undulating Land, 1983), *Byakkō* (White Light, 1988), and *Sakagami* (Sakagami, 1990).

Uno Chiyo (1897–)

Born in 1897 in Iwakuni, Yamaguchi. After graduating from high school she taught briefly at an elementary school. Then, in 1915, she spent a year in Korea. Subsequently she married Fujimura Tadashi and moved to Hokkaido, where she began writing fiction. In 1921 her "Shifun no kao" (The Face with Makeup) received first prize in the *Jiji shimpō* literary contest. *Kōfuku* (Happiness), a collection of her short stories, was published in 1924.

In 1922, Uno moved to Tokyo, divorced Fujimura, and lived with the novelist Ozaki Shirō. Between 1930 and 1935 she lived with Tōgō Seiji, a noted painter, and her major work "Irozange" (The Regret of Love) was written during this time. In 1936 she started a fashion magazine called *Style,* forming a publishing company with the same name, and published a literary journal, *Buntai* (Style), as well.

In 1939 she married Kitahara Takeo, a writer and a scholar of French literature. She began her most well-received work, "Ohan," in 1947, and upon its completion in 1957, she received the Noma Hiroshi Prize. The first and second parts of "Sasu" (To Stab) appeared in the January issue of *Shinchō* in 1963, with the third and the fourth parts appearing in 1965, and the final part in 1966. During that time she was divorced from Kitahara. In 1971 she received the Women's Literature Prize and in 1972, the Art Academy Prize.

Her major works include, in addition to those already mentioned, *Ningyōshi Tenguya Hisakichi* (The Puppetmaker Tenguya Hisakichi, 1943), *Teisetsu* (Chastity, 1970), and *Ame no oto* (The Sound of Rain, 1974). She wrote repeatedly on the themes of love, parting, attachment, and desperation stemming from the relations between man and woman.

She is also a professional kimono designer, and has published numerous essays on kimono. The novel *Usuzumi no sakura* (Cherry Tree in Thin Ink and Water, 1975), narrated by a kimono designer, exemplifies the combining of Uno's two careers. *Uno Chiyo Zenshū* (Complete Works of Uno Chiyo) in twelve volumes was published by Chūō Kōronsha in 1977–78.

About the Translators

Noriko Mizuta Lippit received her Ph.D. from Yale University. She has taught English, American, Japanese, and Comparative literature at Dokkyo University, Josai University, and Tokyo Women's University in Japan, and at Marymount College, Scripps College, and the University of Southern California in the United States. She is currently Director of the Center for Inter-Cultural Studies and Education at Josai University. Her works of criticism include *Reality and Fiction in Modern Japanese Literature* (1980), *A Disturbance in Mirrors: The Poetry of Sylvia Plath* (1981), *Crime and Dream: The World of Edgar Allan Poe* (1982), *Modern Literature and the Female Ego* (1982), and *Beyond Feminism: Modern Literature and Women's Subsconscious* (1991). She is the author of two volumes of poetry, *At the End of Spring* (1976) and *Intervals* (1981), and edited *Women's Self-Expression in Literature, Film and Art* (1991). She is the editor of *Review of Japanese Culture and Society.*

Kyoko Iriye Selden is a graduate of Tokyo University and of Yale University, where she received a Ph.D. in English. She is the coeditor and translator of *The Atomic Bomb, Voices from Hiroshima and Nagasaki* (M.E. Sharpe, 1989), and translator, with Noriko Mizuta Lippit, of *The Short Stories of Tomioka Taeko* (forthcoming). She has taught English, Japanese, and comparative literature at Tsuda College (Tokyo) and Washington University (St. Louis). Currently she teaches Japanese language-literature at Cornell University.